Quick Guide

FENCES & GATES

CREATIVE HOMEOWNER PRESS®

COPYRIGHT © 1993
CREATIVE HOMEOWNER PRESS®
A Division of Federal Marketing Corp.
Upper Saddle River, NJ

Manufactured in the United States of America

Technical Reviewer: Laura Tringali
Cover Design: Warren Ramezzana
Cover Illustrations: Moffit Cecil
Book Packager: Scharff Limited
Electronic Prepress: TSI Graphics
Printed at: Banta Company

Current Printing (last digit)
10 9 8 7 6 5 4 3 2 1

Quick Guide: Fences & Gates
LC: 93-71660
ISBN: 1-880029-22-7

CREATIVE HOMEOWNER PRESS®
A Division of Federal Marketing Corp.
24 Park Way
Upper Saddle River, NJ 07458

C O N T E N T S

S A F E T Y   F I R S T

Though all the designs and methods in this book have been tested for safety, it is not possible to overstate the importance of using the safest construction methods possible. What follows are reminders; some do's and don'ts of basic carpentry. They are not substitutes for your own common sense.

■ *Always* use caution, care, and good judgment when following the procedures described in this book.

■ *Always* be sure that the electrical setup is safe; be sure that no circuit is overloaded, and that all power tools and electrical outlets are properly grounded. Do not use power tools in wet locations.

■ *Always* read container labels on paints, solvents, and other products. Provide proper ventilation, and observe all other warnings.

■ *Always* read the tool manufacturer's instructions for using a tool, especially the warnings.

■ *Always* use holders or pushers to work pieces shorter than 3 inches on a table saw or jointer. Avoid working short pieces if you can.

■ *Always* remove the key from any drill chuck (portable or press) before starting the drill.

■ *Always* pay deliberate attention to how a tool works so that you can avoid being injured.

■ *Always* know the limitations of your tools. Do not try to force them to do what they were not designed to do.

■ *Always* make sure that any adjustment is locked before proceeding. For example, always check the rip fence on a table saw or the bevel adjustment on a portable saw before start working on a project.

■ *Always* clamp small pieces firmly to a bench or other work surface when sawing or drilling.

■ *Always* wear the appropriate rubber or work gloves when handling chemicals, during heavy construction, or when sanding.

■ *Always* wear a disposable mask when working with odors, dusts, or mists. Use a special respirator when working with toxic substances.

■ *Always* wear eye protection, especially when using power tools or striking metal on metal or metal on concrete; a chip can fly off, for example, when chiseling concrete.

■ *Always* be aware that there is never time for your body's reflexes to save you from injury from a power tool in a dangerous situation. Everything happens too fast—be *alert!*

■ *Always* keep your hands away from the business ends of blades, cutters, and bits.

■ *Always* hold a portable circular saw with both hands so that you will know where your hands are.

■ *Always* use a drill with an auxiliary handle to control the torque when large bits are used.

■ *Always* check your local building codes when planning new construction. The codes are intended to protect public safety—you should observe them to the letter.

■ *Never* work with power tools when you are tired or under the influence of alcohol or drugs.

■ *Never* cut very small pieces of wood or pipe. Whenever possible, cut small pieces off larger pieces.

■ *Never* change a blade or a bit unless the power cord is unplugged. Do not depend on the switch being off; you might accidentally hit it.

■ *Never* work in insufficient lighting.

■ *Never* work while wearing loose clothing, hanging hair, open cuffs, or jewelry.

■ *Never* work with dull tools. Have them professionally sharpened, or learn how to sharpen them yourself, using proper equipment.

■ *Never* use a power tool on a workpiece that is not firmly supported or clamped.

■ *Never* saw a workpiece that spans a large distance between horses without close support on either side of the kerf; the piece can bend, closing the kerf and jamming the blade, causing saw kickback.

■ *Never* support a workpiece with your leg or other part of your body when sawing.

■ *Never* carry sharp or pointed tools, such as utility knives, awls, or chisels in your pocket. If you want to carry tools, use a special-purpose tool belt with leather pockets and holders.

DESIGN CHECKLIST

A fence can be as vital a part of your home's exterior as flowers, shrubbery, and trees. Before you start building a fence, undertake a thorough investigation of the elements that will influence the type and size of the fence. There are a few important questions to consider.

What function do you want the fence to perform?

What aspect of the fence is most important: utility or appearance?

How is the fence going to mesh with the existing landscape?

Functional Considerations

The fence has long been used to serve two main functions: define boundary lines and provide security. Over the years, homeowners have discovered many more roles that fences can play in residential living. By answering the questions "What do I want my fence to do for me?" you are well on your way to discovering a successful design solution. Do you want to separate areas of outdoor work, play, and storage? Do you want to provide a measure of backyard privacy? Do you want to keep your youngsters away from the road or a body of water? Do you want to prevent the neighborhood kids from using your yard as a shortcut? Do you want to protect a cherished garden from trampling feet? Do you want shade for a patio? Do you want to allow some sunshine in, while still protecting against the wind? Once you have considered these questions carefully, you can more intelligently choose the type of fence that will best serve your needs.

Privacy. Today, privacy is of utmost importance to many homeowners in their selection of fences. The degree of privacy desired will influence your choice of materials. For example, plywood siding, solid boards, close-set grapestakes, and basketweave designs provide maximum screening and privacy. Partial privacy can be achieved through the use of vertical louvers, slats, and vertical board and board fences. Fencing materials that offer little or no privacy are lattice, chain link, bamboo, and post and rail. The most obvious way to achieve total privacy is to rim the property with a high solid fence. Once the fence is up, however, you may find that you have created a rigid environment with a confined, boarded-up feeling. In such cases, plantings can help transform an unfriendly fence into an attractive addition to the landscape.

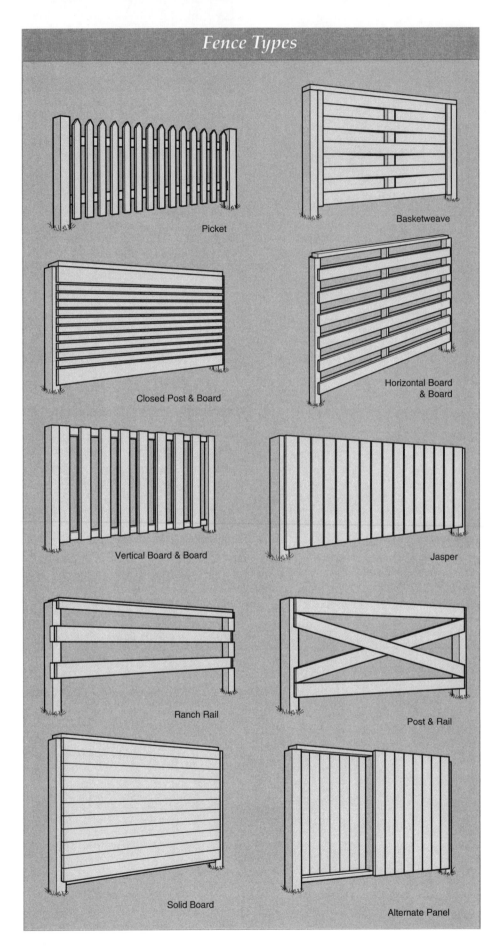

Fence Types

Picket

Basketweave

Closed Post & Board

Horizontal Board & Board

Vertical Board & Board

Jasper

Ranch Rail

Post & Rail

Solid Board

Alternate Panel

Alternate Horizontal & Vertical

Alternate Width

Dog-Ear Privacy

Louver

Lattice Top

Bamboo

Diagonal Board

Curved Panel Picket

Grapestake

Split Rail

Mortised Rail

Palisade

Horizontal Louver

Plywood Siding

Aluminum Panel

Chain Link

Security. The role of security is more important than ever in residential fencing. To keep out unwelcome guests, a high chain link or plywood siding fence is probably the best choice. However, security also can mean keeping small children or animals within a certain area. In these instances, the fence needs to be just high enough to thwart a climbing child or jumping dog. A good choice would be the closed post and board.

Protection against Sun. Heat and glare from the sun can be controlled through thoughtful selection and placement of fencing. For dappled shade, sunlight can be filtered through a louvered, basketweave, or slat fence. A solid board fence around a patio will keep out the late afternoon sun that slants in under trees and roof overhangs. This type of fence can also be used to cool the western wall of a house by blocking the sun before it reaches the walls and windows.

Boundaries. Boundary fences are basically territorial markers and can be among the cheapest fences to build in terms of materials and labor. They need only be short, lightweight structures that point out the limits of your property. Naturally, proper consideration must be given to zoning restrictions and actual property lines. Post-and-rail and ranch-rail fences are popular choices for boundary fences.

Reducing Noise. Noise is a common complaint, especially in urban areas. Homeowners near freeways and highly traveled streets often wonder if there is something they can do with fences to reduce traffic sounds.

Certain fences are effective in reducing noise levels, but only to a degree. To be effective, a fence must be solid and heavy (thickness and mass are important). It should also extend several feet above the source of the sound and the receiver— the higher the better. Good choices for noise reduction would be the alternate panel and jasper fences.

The Question of Responsibility

Problems can arise when your fence is to be built between adjoining properties. Fences are generally considered as belonging to the land on which they are built, and a line or division fence belongs to the neighboring property owners as tenants in common. An accurately surveyed boundary is necessary to ensure that the fence is being put where it is intended to be. It is also a good idea to check the deed and title of your property. Not only will this spell out the exact boundaries of your property, but it will also inform you of any restrictions that may have been placed on it by previous owners.

It is desirable, if possible, to have an agreement with the adjoining property's landowner in advance. This agreement should determine the location and type of fence and divide up the responsibility for building and maintaining it. Such agreements, which should be in writing, can avoid many disputes later on. They can be recorded in most states with the effect of fixing the responsibility for fence maintenance between future purchasers of the property.

Boundary fences erected by adjoining property landowners together are usually placed on the property line, partly on the property of each, and are commonly owned. However, it is not always possible to get the agreement and help of a neighbor in erecting a fence. In such cases, it is advisable to place the fence entirely on the land of the builder, or a few inches inside the line to allow for possible errors in the survey. Such a fence is entirely the property of the landowner, who does not need to consult with the neighbor regarding its erection or maintenance as long as it does not violate any law or ordinance.

Unfortunately, division fences provide many opportunities for disputes. The solution of such disagreements by force of law is not always satisfactory because neighbors must go on living next to each other afterward. A reasonable amount of cooperation in the building and maintaining of fences for the mutual satisfaction of all can help you and your neighbor avoid or settle most problems.

Laying Out a Curved Fence

To lay out a fence on a contour, start by staking out a smooth curve, placing the stakes approximately 16 feet apart. If a sharp curve presents itself, reduce the stake spacing to 8 feet; place three stakes at the sharpest point of the curve. Now stretch a string between the two end stakes and measure the setback of the center stake from the string. If the setback is 10 inches or less, the normal 8-foot post spacing is acceptable. For every 2 inches of setback over 10, decrease the post spacing by 1 foot. By reducing the post spacing and keeping the curve smooth, the fence will pull evenly on all the posts.

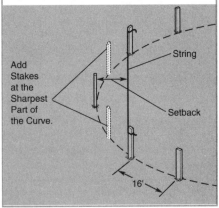

Add Stakes at the Sharpest Part of the Curve.

String

Setback

16'

Much like a patio or deck, fencing can be used to transform unused living space into attractive outdoor rooms. Short, free-standing fences can effectively separate the outdoor living area from the service area, gardening center, and play yard, thus providing separate areas for work, play, and relaxation. In creating an outdoor room, the trees, sky, and horizon beyond are elements that must be considered. Some may be screened out or incorporated into your plan. Placement of a narrow fence in a specific place may be all that is needed to blot out an objectionable view. If the land slopes down to a busy street or freeway, a long, low fence can conceal the road below, yet provide an unobstructed view of the hills beyond. A large, open yard can be made more intimate if it is broken up with fencing.

Whatever the type or style of fence you ultimately choose, it should relate to your home's exterior, not stand out as an apparent afterthought. Choose only those styles and materials that will harmonize with your home and the entire landscape design.

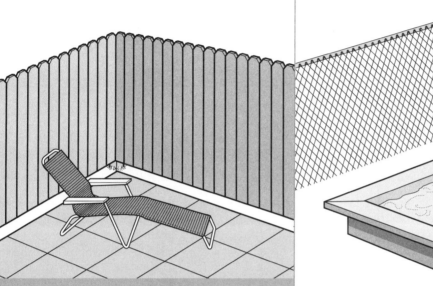

The desire for patio privacy is reflected in the use of this dog-ear fence.

The safety of children is a major reason for the erection of the chain link and other security fences.

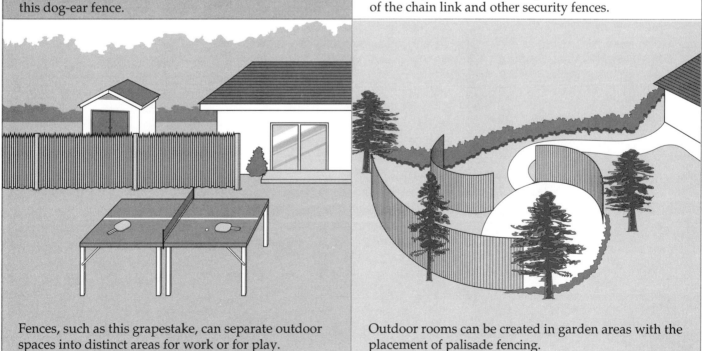

Fences, such as this grapestake, can separate outdoor spaces into distinct areas for work or for play.

Outdoor rooms can be created in garden areas with the placement of palisade fencing.

Site Considerations

When you have clearly identified what functions you want your fence to fulfill, the next step is to understand the physical and environmental limitations of your site. These restrictions can influence the physical design of the fence. It is essential that you evaluate the following considerations.

Terrain. The location of your fence and the type you choose to build may be dictated by the terrain. If your property were perfectly flat, selecting and laying out a fence would be simple. In most cases, however, you will probably have to deal with at least some uneven terrain. Before proceeding with the design and layout, you should first determine the slope of your property where the fence will travel.

If fencing along sloped terrain, you can either lay out the fence in steps or follow the natural contours of the land. In general, fences on short, steep slopes look better when they are stepped; if the fence follows the slope, it appears to bulge or lean. Additional framing might be required on very steep slopes where you step the fence down so far that the distance between the bottom rail and the ground is excessive. Longer or gentler slopes can be followed. Some types of fencing, such as the post and rail, lend themselves easily to hillside contours. Less flexible are the more geometric forms, such as solid board and louver; these fences are best stepped. Make a sketch of the slope to scale, draw the fence both stepped along the slope and with the fence following the slope, then decide which you think looks best.

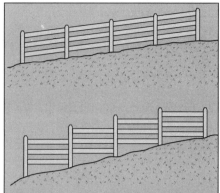

Terrain. On gentle slopes, the fence follows the slope (top). On steep slopes, the fence is stepped accordingly (bottom).

Calculating the Slope

To determine the slope, put a 2x2 stake in the ground at the top of the slope. Put another 2x2 stake in level ground at the bottom of the slope. This second stake has to be taller than the slope. Stretch a string taut between the stakes and level it with a line level hung at the center of the string. To ensure an accurate reading, take the measurements on a windless day and make sure the string does not touch anything—not even a few blades of grass. Measure the distance from the string on the taller stake to the ground. This is the rise of the slope. Divide the rise by the run (the length of line between stakes) to find the slope.

Use graph paper to draw a sketch of the slope to scale. A convenient scale to use is 1 square = 2 inches. Sketch the profile of the slope accurately. Then lay tracing paper over the profile and draw different fence designs, both following the slope and stepping down it.

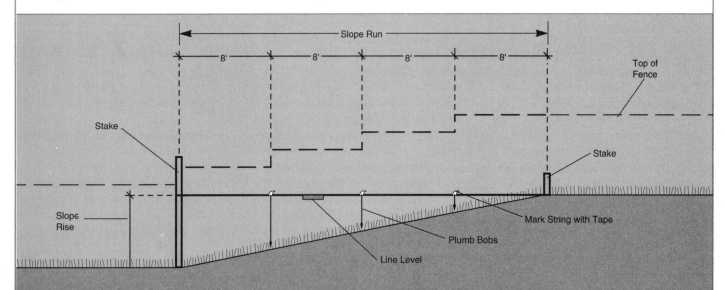

Plotting the rise and run on sloping ground is not difficult, but it requires careful measurements. The proportional relationship is determined by the width of the fence section; in this case, 8 feet.

Frost Heave. Most types of soil are sufficiently stable to support a fence, but cold climates create special problems—especially for fences built on clay soil. Water that freezes in any soil will cause it to expand, often moving or damaging structures within the soil, but clay soils are the worst. For example, in a very cold region, a fence post that was dropped into moist spring clay may be squirted out the following January. This natural process is called frost heave.

There is no cure for frost heave, but there are a few ways to minimize its effects:

■ Improve drainage beneath and around underground structural elements so that water cannot build up. This is accomplished by removing some of the clay soil and replacing it with sand or gravel. However, this solution may be impractical in areas where the frost line is almost as deep as the aboveground height of the fence.

■ Another way of protecting structures beneath the ground is to extend the footings beneath the frost line. This method can be expensive in areas of extreme cold.

For protection against frost heave in those areas of the country where it is a serious problem, you need to consult either a local civil engineer or your building department for advice pertaining to your particular project.

Drainage is a complicated issue. Professionals, laymen, and others associated with building and landscaping have developed standards for building and planting in all parts of the country. Often a phone call to the right person (your building department or agricultural extension service is the place to start) can give you all the information you need in a few minutes. Usually they will assist you without assessing a fee.

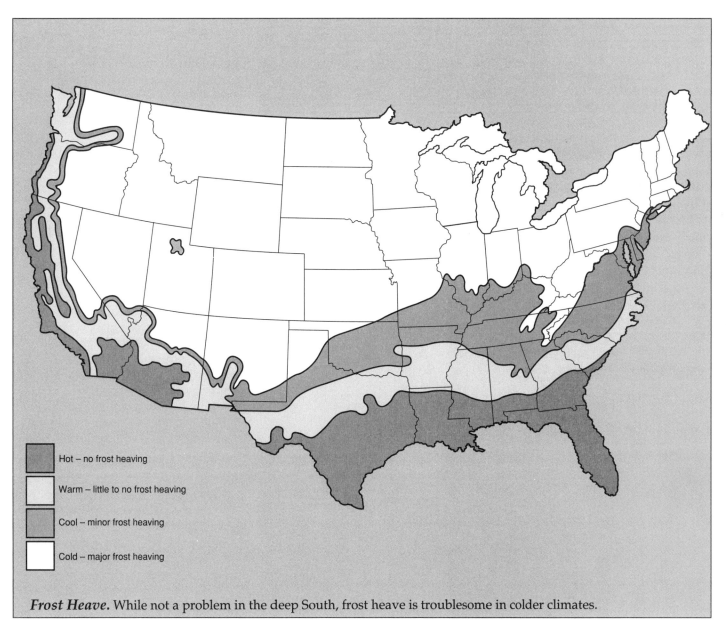

Hot – no frost heaving

Warm – little to no frost heaving

Cool – minor frost heaving

Cold – major frost heaving

Frost Heave. While not a problem in the deep South, frost heave is troublesome in colder climates.

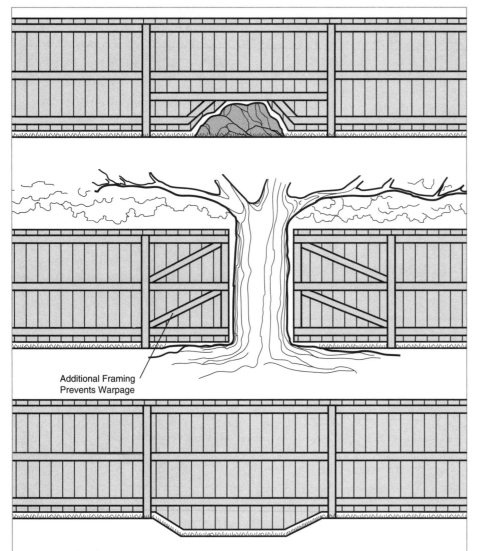

Additional Framing
Prevents Warpage

Natural Obstructions. Fence sections may be adjusted to allow for natural obstructions that you cannot or do not wish to remove. Do not extend the boards more than a few inches below the bottom rail or they will be prone to warpage.

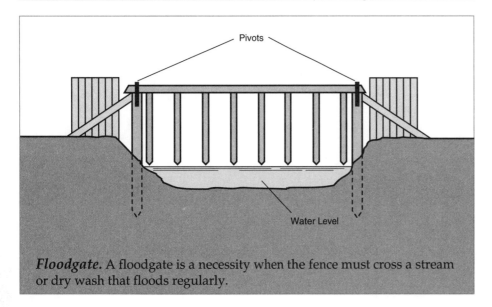

Pivots

Water Level

Floodgate. A floodgate is a necessity when the fence must cross a stream or dry wash that floods regularly.

Natural Obstructions. There are times when a fence must be run above an immovable object (a boulder, for example) or dropped lower than its normal bottom to fill a void in the ground. Sometimes erecting a fence is complicated by one or more trees growing right on the line where the fence should go. If the tree is small, your best solution may be to remove it. However, if it is a large specimen that you want to preserve, you can incorporate it into the fence. Stop the fence an inch or two short of the trunk, and be sure not to place the last post so close that you will injure the root system when you dig the posthole. The fence should also be designed with additional framing to prevent warping. Never plan to include a tree in your fence line as a post. The chief hazard in using the tree for a post is the possible injury that may be done to the tree itself. If you must attach anything to the trunk, be sparing in the use of nails.

Some ingenuity is required when a fence must cross a stream or dry wash that floods in the spring. The risk is that you will create a barrier that will become choked with debris; the impeded waterflow will either cause a neighborhood flood or wash out the fence. One way to solve the problem is to install a floodgate, a simple, rugged device that opens automatically at the flood stage. A typical floodgate pivots on anchor posts on each side of the stream and swings upward as the water rises. Two vital requirements are strong anchor posts on each bank and measures to prevent erosion of the bank, such as concrete side walls. Gates should be cleaned out after every passage of floodwater.

When the fence has to be built along the edge of a stream or bluff, the layout should be worked out with a landscape architect. The architect can tell you how close to the edge the fence can be without losing it eventually to the wind or water action. Often, plantings along the edge will slow down the soil movement.

Wind Conditions. Like airplanes and automobiles, fences are subject to some rather strange principles of aerodynamics. While it might seem that a perfectly solid fence would make the best windbreak, this is surprisingly not the case. In fact, this type of fence actually aggravates the wind problem by bringing the currents down to ground level. In addition, a severe storm could flatten the entire fence.

In most cases, it is not necessary to completely stop the wind. If the winds in your area blow strong and often, the goal is to slow them down as much as possible to lessen their

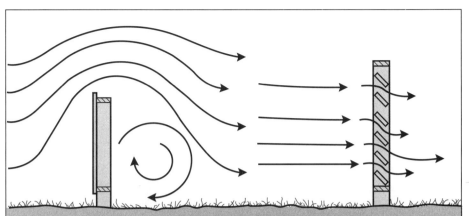

Wind Conditions. Wind that strikes a solid fence tends to "hop" over it, creating a vacuum on the opposite side that pulls the air down again (left). A louvered fence will not totally block the wind, but will slow its velocity to a more comfortable level (right).

Zoning & Code Requirements

Zoning. Zoning laws control land use and the density of building in prescribed areas. They specify setback requirements, which state how close you can build to any property line, the height of fences and trellises, and in some cases, what materials may be used. They may specify how much of your lot may be covered by structures. Building departments often publish data sheets that answer questions about residential landscaping; these are available for you to consult.

If your plans conflict with regulations, you can apply for a variance, that is, a permit to build a structure that does not conform in detail to the law. Do not put up a fence that violates your local zoning laws and codes. A building inspector's report could lead to a legal order to tear it down.

Determine whether there are any easements spelled out in the property deed. An easement is a right-of-way granted to a utility company or other property owner. Your deed may include other stipulations that limit

the design or the location of a new structure. Check your deed before you build.

Codes. Building codes are intended to protect public safety. Before beginning any fence construction, contact your local authorities for building code requirements in your community. Do not forget to ask if a permit is required for your project. Take the time to meet your local code and permit requirements and you are well on your way to a successful fence project.

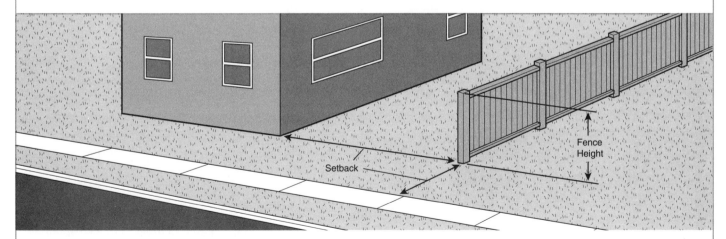

Local building codes may restrict the locations and dimensions of fences in relation to the street or to your next-door neighbors.

effect. One way of doing this is by building fences with openwork screening such as lattice and slats. Louvers not only reduce the air's velocity considerably, but they redirect it as well. In fact, the louvered fence has been found to give the best protection over the greatest distance. Horizontal louvers that have been directed down can even be used to catch cool evening breezes in the summertime.

Before you locate a windbreak fence, contact the local weather bureau for the direction and severity of prevailing winds. If you happen to live in a rural area, ask your local agricultural extension office for advice.

Complementing Your Home

Unless you are building your fence in an isolated area, it is important that you consider the architecture, materials, and decor of your house. If the materials of both house and fence are compatible, this will produce a natural transition between the two. The fence should complement your home's decor, not clash with it. For instance, rural settings tend toward wood fences—a weathered post and rail for a classical look, a picket or lattice top for a cottage. A modern city dwelling is better suited to ornamental aluminum; if a wood fence is preferred, the formal look of a solid board is a good choice.

The major criteria for selecting fence styles and materials are these: They should be in keeping with the surrounding architecture. They should be available locally in pleasing colors and texture. They should require low maintenance and be economically feasible. By following these criteria and using a little imagination, your new fence can do more than just fill a need—it can enhance your home's exterior.

Wood is the most popular fence material. The visual qualities of wood—its grain, texture, and color—make it an exciting material to work with. Lumber is sold in different grades and with different surface treatments. The best grade is clear, meaning that it is free of knots and other blemishes. The next grade is select; which is divided into #1, #2, and #3. No. 1 select has minimal blemishes and as the numbers get higher the blemishes increase, while the cost of the wood decreases. The lowest grade, though still perfectly acceptable, is common. This wood will have more than a few blemishes. For economy, choose the lowest grade that will be adequate for your fence design. If you will be painting or staining the fence, a lower grade can be used; the finish will hide most of the blemishes.

Popular lumber choices include redwood, cypress, and spruce, all of which possess excellent weathering durability. If left unpainted, these woods gradually acquire an attractive, soft gray patina. Some experts recommend using only heartwood for fence posts. Heartwood comes from the center of the tree and is denser and more durable than the surrounding sapwood. Regardless of the type and grade of lumber used, make sure it is free of rot and insect infestation, and avoid wood that has warped.

A wood choice that is gaining in popularity is pressure-treated lumber. This wood is highly resistant to decay caused by the elements and insects, so it is ideal for fences. Some pressure-treated products also contain a colorant that causes the wood to weather to a pleasing, mellow gray similar to untreated redwood. Pressure-treated lumber readily accepts paints and stains.

This picket fence, complete with ornamental tops, is a perfect match for a serene country cottage.

This solid board fence makes a good match with the contemporary architecture of a Ranch-style house.

FENCE BUILDING BASICS

Before starting the construction of a fence, identify all those pieces that will make up the finished fence. Almost every type of fence consists of two major structural parts. Posts are the vertical framing members set in the ground, and rails are the horizontal framing members to which the fence boards are attached.

The posts are the foundation of the fence. Because they are sunk in the ground, wood posts must be chemically treated to prevent decay. Fence posts are typically 4x4 or 6x6 lumber. These sizes are good for most allowable fence heights with typical fence boards. Rails are at least 2x4 and perhaps 2x6 or larger, depending on the distance between the posts and the weight of the fence boards.

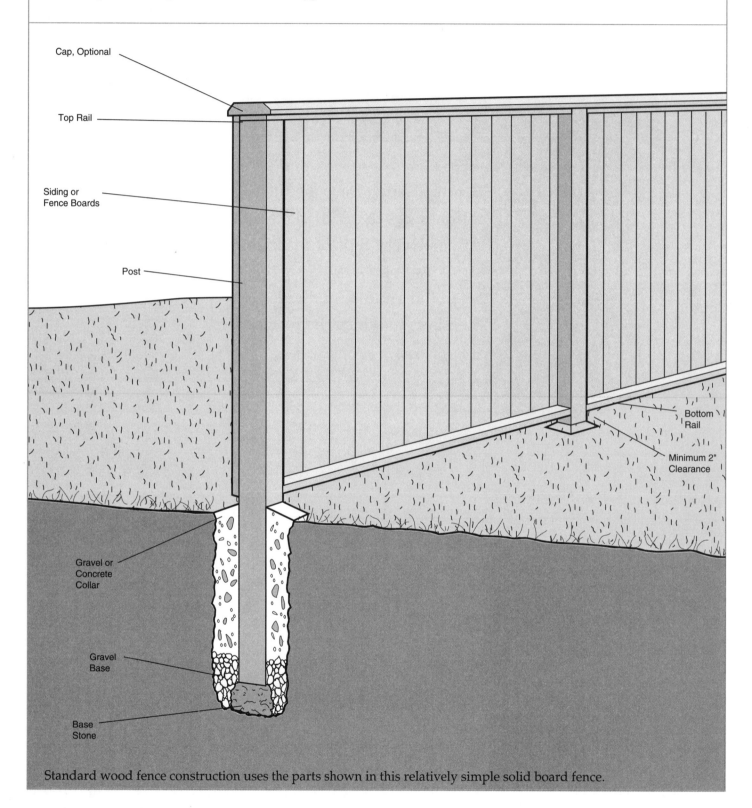

Cap, Optional

Top Rail

Siding or Fence Boards

Post

Bottom Rail

Minimum 2" Clearance

Gravel or Concrete Collar

Gravel Base

Base Stone

Standard wood fence construction uses the parts shown in this relatively simple solid board fence.

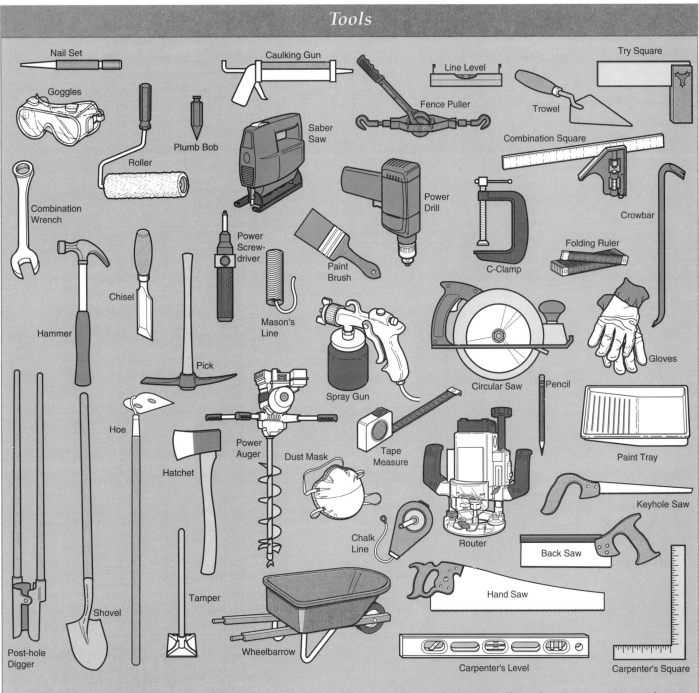

Nail Set
Goggles
Roller
Plumb Bob
Combination Wrench
Hammer
Chisel
Pick
Hoe
Hatchet
Power Auger
Tamper
Shovel
Post-hole Digger
Caulking Gun
Saber Saw
Power Screwdriver
Mason's Line
Paint Brush
Power Drill
Spray Gun
Dust Mask
Wheelbarrow
Line Level
Fence Puller
Circular Saw
Chalk Line
Tape Measure
Router
Hand Saw
Carpenter's Level
Try Square
Trowel
Combination Square
C-Clamp
Crowbar
Folding Ruler
Gloves
Pencil
Paint Tray
Keyhole Saw
Back Saw
Carpenter's Square

The tools and equipment shown above are essential for preparing the postholes, installing the posts, cutting and assembling the parts, and maintaining the fences described in this book.

Tools

You can build a fence with a modest set of tools (see page 23). To dig the postholes, you will need a post-hole digger and possibly a pick and shovel. While the manual clamshell-type post-hole digger is more than adequate, power augers make quick work of digging. If you have a lot of holes to dig, you may wish to rent a power auger. For mixing concrete, you will need a wheelbarrow, hoe, and shovel. For setting out the plan, have a tape measure, folding ruler, mason's line, pencil, plumb bob, chalk line, line level, carpenter's level, and carpenter's square. To cut wood, you will need a hand saw or power saw. For assembling the fence, a power drill, hammer, power screwdriver, combination wrenches for carriage bolts, nail set, and chisel are necessary. When finishing the surface, brushes, a paint tray, and perhaps a roller will be needed. Be sure you have the proper safety equipment: goggles, dust mask, and gloves.

Fasteners

All fasteners and hardware must be specifically intended for outdoor use and be either hot-dipped galvanized or aluminum. When corrosion problems run to extremes, such as on or around salt waterand in high-acid environments, stainless steel is a good choice. The purpose of using these types of fasteners is to prevent harmful corrosion from attacking the wood. Fasteners made of cheaper materials will quickly rust

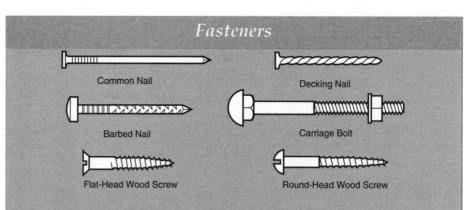

Fasteners

Common Nail

Decking Nail

Barbed Nail

Carriage Bolt

Flat-Head Wood Screw

Round-Head Wood Screw

Each fastener shown above has its individual function in fence construction; when used for that purpose, it ensures a reliable, long-lasting structure.

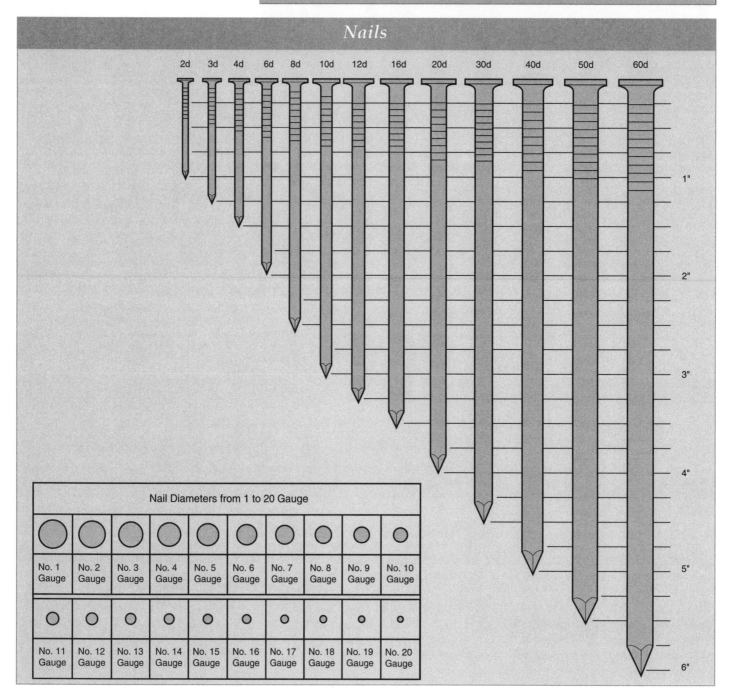

Nails

2d 3d 4d 6d 8d 10d 12d 16d 20d 30d 40d 50d 60d

1"
2"
3"
4"
5"
6"

Nail Diameters from 1 to 20 Gauge

No. 1 Gauge	No. 2 Gauge	No. 3 Gauge	No. 4 Gauge	No. 5 Gauge	No. 6 Gauge	No. 7 Gauge	No. 8 Gauge	No. 9 Gauge	No. 10 Gauge
No. 11 Gauge	No. 12 Gauge	No. 13 Gauge	No. 14 Gauge	No. 15 Gauge	No. 16 Gauge	No. 17 Gauge	No. 18 Gauge	No. 19 Gauge	No. 20 Gauge

and spoil the job. Fasteners used in fence construction are shown on page 18 and are described here.

Nails. Galvanized nails are readily available and will not rust as long as the zinc coating is not broken. They can be used for all parts of your fence construction. Aluminum nails resist rust better than galvanized nails, but they are not as strong and cost more. Stainless steel nails are available in some places; they resist rust best of all, but are the most expensive. Nails are usually sold by the pound, either from open stock or in packages. The price depends on their material, coating (if any), design, and size. The length of a nail is indicated by its penny size; the letter designation for the word penny is the "d."

While common nails provide adequate gripping, there are better choices for optimum holding power. Decking nails have a spiral or screw-type shank to resist loosening or popping out over time. Barbed nails also provide very good resistance to pull-out forces; they are often used to install fence brackets and T-plates.

Nails used to attach fence boards to posts should be three times longer than the thickness of the board itself. Thus, a nominal 1-inch board, which is actually 3/4 inch thick, will require a nail 2 1/4 inches long. An actual 1-inch thick board calls for a 3-inch long nail.

Nails should always be placed carefully so as to provide the greatest holding power. Placing nails too close to the edges of the wood will produce splitting. Avoid nailing into the end grain; if you must fasten into the end grain, use screws instead. When driving more than one nail, always stagger the nails so that none are in the same grain line. A few nails of the proper type and size, properly placed and driven, will hold better than a great many nails driven close together.

Screws and Bolts. These fasteners are more expensive than nails and take longer to install, but the resulting connection is stronger.

Steel wood screws are generally the strongest, but unless they are specially plated to resist rust, brass screws are better for use outdoors. Chrome- or nickel-plated screws are also ideal for exterior use. Flat head screws are used when you want the head below the surface of the wood; round head screws protrude above the surface. The threaded part of the screw should reach completely into the second piece of wood. First drill pilot holes to prevent the wood from splitting. A screw-pilot bit in a power drill produces both a pilot hole for the screw threads and a countersink hole for the screw head.

Carriage bolts provide a neat, finished appearance and superior strength. Buy bolts that are 1 inch longer than the total thickness of the pieces being joined. Drill a hole through the wood that is 1/16 inch larger than the bolt diameter.

Screws

LENGTH	GAUGE NUMBERS																	
	#0	#1	#2	#3	#4	#5	#6	#7	#8	#9	#10	#11	#12	#14	#16	#18	#20	#24
1/4 inch	#0	#1	#2	#3														
3/8 inch			#2	#3	#4	#5	#6	#7										
1/2 inch			#2	#3	#4	#5	#6	#7	#8									
5/8 inch				#3	#4	#5	#6	#7	#8	#9	#10							
3/4 inch					#4	#5	#6	#7	#8	#9	#10	#11						
7/8 inch							#6	#7	#8	#9	#10	#11	#12					
1 inch							#6	#7	#8	#9	#10	#11	#12	#14				
1 1/4 inches								#7	#8	#9	#10	#11	#12	#14	#16			
1 1/2 inches							#6	#7	#8	#9	#10	#11	#12	#14	#16	#18		
1 3/4 inches									#8	#9	#10	#11	#12	#14	#16	#18	#20	
2 inches									#8	#9	#10	#11	#12	#14	#16	#18	#20	
2 1/4 inches										#9	#10	#11	#12	#14	#16	#18	#20	
2 1/2 inches													#12	#14	#16	#18	#20	
2 3/4 inches														#14	#16	#18	#20	
3 inches															#16	#18	#20	
3 1/2 inches																#18	#20	#24
4 inches																#18	#20	#24

Connectors

When used in conjunction with fasteners, metal connectors add rigidity to a fence and ensure stronger and longer-lasting connections (see page 25). Fence brackets and T-plates provide a quick and easy method of connecting the rails to the posts without the need for special cutting, notching, or other fitting.

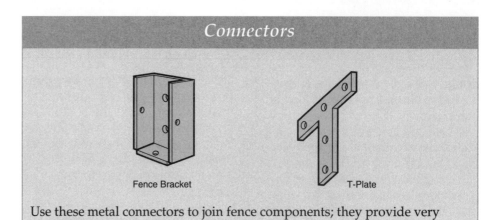

Connectors

Fence Bracket

T-Plate

Use these metal connectors to join fence components; they provide very stable fences.

Basic Procedures for Using Fasteners

Blunted Tip

Tap gently with a hammer.

Blunt the ends of sharp nails with a hammer to avoid splitting the fence boards

Incorrect

Correct

To avoid splitting the wood when nailing, stagger the nails so none are in the same grain line.

The threads of a wood screw should be completely sunk into the second piece of wood.

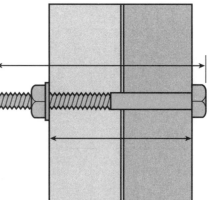

Use bolts that are 1 inch longer than the total thickness of the boards being joined.

In addition to using connecting hardware, there are some basic woodworking joints that can be used for attaching rails to posts. The joint you use will depend largely on the type of fence you are building. Three of the most popular woodworking joints are the butt, lap, and dado.

Butt Joint. This joint is relatively easy to make. The end of the rail is placed against the post and nailed in position. Butt joints are rather weak and often require support blocks, or nailers, underneath to give additional strength.

Lap Joint. This is also an easy joint to make. It requires nailing the rail to the side or top of the post and butting the ends of rails together. The lap joint can be a top lap or side lap. A variation on the top lap features mitered rail ends at the corners.

Dado Joint. A dado joint is similar to the lap joint, except the rails are grooved or set into the post, instead of being nailed to the surface. Because dado joints are very good weight bearers, they are especially well-suited for heavy board fences.

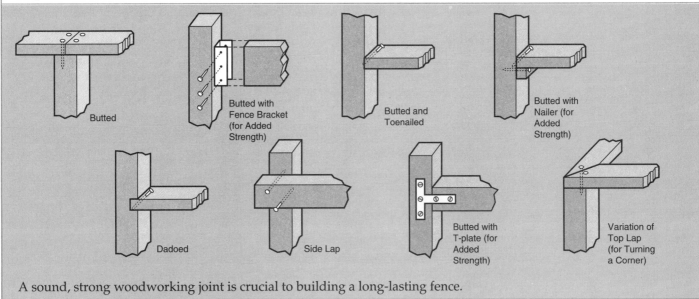

Butted

Butted with Fence Bracket (for Added Strength)

Butted and Toenailed

Butted with Nailer (for Added Strength)

Dadoed

Side Lap

Butted with T-plate (for Added Strength)

Variation of Top Lap (for Turning a Corner)

A sound, strong woodworking joint is crucial to building a long-lasting fence.

Cutting a Dado

Dadoes are channels cut across the grain of a board, into which a second piece of wood is fitted. The most common is the housed dado, in which the entire width of the second ("housed") piece fits into the channel, or dado. A variation is the stopped housed dado, or combination dado and rabbet. In this version, the dado is cut across the full width of the piece, with a rabbet cut on the housed member.

While a chisel and handsaw can be used for this job, it is much easier to cut a dado with a router. There are many router bits available to cut whatever width and depth you wish. Use C-clamps to secure a straightedge in line with the cut you want to make. The straightedge will act as a guide for the base plate of the router. Clamp the straightedge so that the cutter will make the dado in the exact spot you wish. You can use a ruler and carpenter's square to measure and mark the dimensions.

No matter how you make the groove for the dado, make sure the cut is smooth and the edges sharp so that it can readily accept the mating piece of wood. A piece of tightly folded sandpaper usually works to clean the cut after it has been made. If the cut is very rough, go over it again with the router.

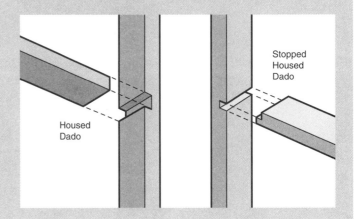

Stopped Housed Dado

Housed Dado

When you have decided on a particular type of fence, you must estimate the amount of wood and other materials required to build it. To determine the linear footage of the fence, measure along the line to be fenced. Do not forget to allow space or any gates.

After measuring the fence's length, divide the length into intervals to determine proper post spacing. Intervals of 6 or 8 feet are the most common. (Prefabricated fencing comes in 6- and 8-foot sections.) Do not place posts more than 8 feet on center because the rails can sag.

Intervals of 6 or 8 feet on center allow you to use common sizes of precut lumber in even lengths. Any odd space left at the end can be used for a gate or a shorter fence section. Note: "On center" relates to the fact that when setting posts you should measure from the center of one post to the center of the next; this measurement should be 6 or 8 feet. To figure the total number of posts, simply add one to the total number of fence sections (nine sections = ten posts, eleven sections = twelve posts, etc.). After you know the quantities you need, you can proceed with cost estimates and ordering.

As a simple method of determining the concrete cost, count on the equivalent of one 80-pound bag of premixed concrete per post. You can adjust up or down from that, depending on how you finally decide to prepare the concrete and how much you need.

An alternative is to divide the fence into equal intervals. This will prevent you from ending up with a short interval at the end of the fence. However, it requires more measuring, cutting, and fitting, which can waste a fair amount of lumber. The advantage to cutting the lumber yourself as you go is that it allows you to make small adjustments in the overall length or height of the fence.

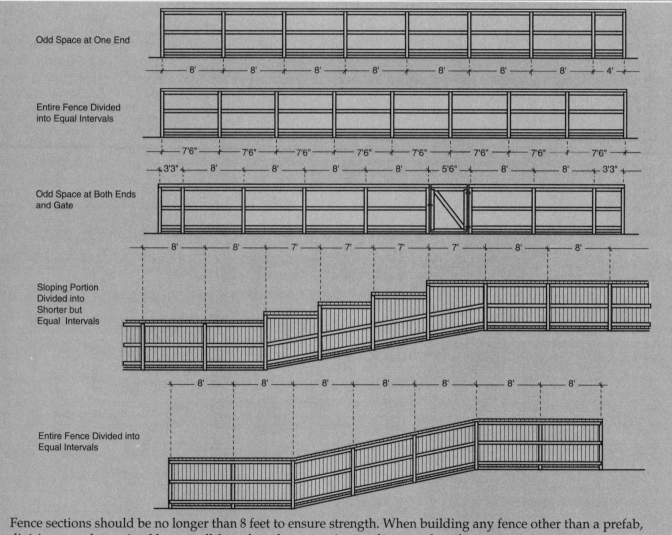

Odd Space at One End
8' 8' 8' 8' 8' 8' 8' 4'

Entire Fence Divided into Equal Intervals
7'6" 7'6" 7'6" 7'6" 7'6" 7'6" 7'6" 7'6"

Odd Space at Both Ends and Gate
3'3" 8' 8' 8' 8' 5'6" 8' 8' 3'3"

Sloping Portion Divided into Shorter but Equal Intervals
8' 8' 7' 7' 7' 7' 8' 8'

Entire Fence Divided into Equal Intervals
8' 8' 8' 8' 8' 8' 8'

Fence sections should be no longer than 8 feet to ensure strength. When building any fence other than a prefab, divisions are determined by overall fence length, proportion, and personal preference.

Build the Basic Fence

It is very important that you work carefully and allow plenty of time for laying out the fence. Remember that a measurement that may be just an inch off at the beginning can turn into feet down the line. Unless the fence is unusually long, it is best to set all the posts first, then add the rails and fence boards later.

1 **Staking Out the Fence.** Locate the fence corners (if any) and ends. Mark them with stakes. For stakes, use 1x2s or 2x2s that have been pointed with a hatchet.

2 **Measuring by Triangulation.** When a fence must turn a corner, square the corner using the 3-4-5 triangular method. The square of the hypotenuse must equal the sum of the squares of the sides.

3 **Locating the Posts.** Measure and locate positions for all posts and mark these locations with stakes, using the string as a guide. Work carefully and check all measurements twice. The structural strength and finished appearance of the fence depend on accurate positioning of the posts.

4 **Laying Out a Curved Fence.** To lay out a curved fence, stake the ends and stretch a line between them. Find the midpoint and, using a homemade line-and-pencil compass, scribe the curved line.

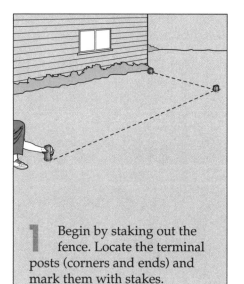

1 Begin by staking out the fence. Locate the terminal posts (corners and ends) and mark them with stakes.

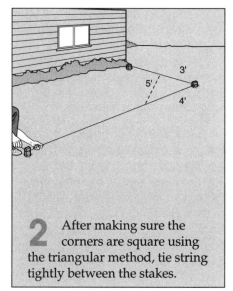

2 After making sure the corners are square using the triangular method, tie string tightly between the stakes.

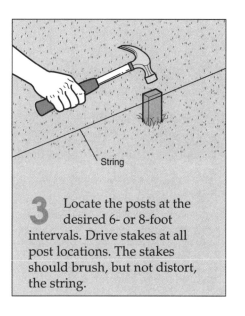

3 Locate the posts at the desired 6- or 8-foot intervals. Drive stakes at all post locations. The stakes should brush, but not distort, the string.

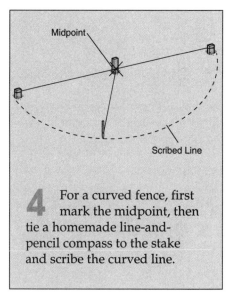

4 For a curved fence, first mark the midpoint, then tie a homemade line-and-pencil compass to the stake and scribe the curved line.

Digging Postholes

Post-hole depth is determined by the height of the fence and the pressure that will be placed on it. Local ordinances also come into play. Many municipalities, for example, have 6-foot-height limits for fences. In any case, fence posts must be sunk into the ground a minimum of 3 feet, though 4 feet is preferable. A 6-foot-fence will use 9- or 10-foot posts sunk 3 or 4 feet,

a 4-foot fence will use 7- or 8-foot posts sunk 3 or 4 feet, and so forth. Even fences less than 4 feet high should have posts at least 3 to 4 feet deep. When you have to buy longer pieces of lumber than you actually need, you might as well put the extra length in the ground instead of cutting it away.

The clamshell-type post-hole digger is probably the best all-around digger.

It is surprisingly fast. For less labor intensive digging, use a power auger. If the soil is very hard or rocky, you might have to use a pick and shovel. In extreme cases, use a jackhammer. Dig postholes a few inches deeper than the depth needed to allow for the footings. Holes should be 10 to 12 inches in diameter for a 4-inch post.

5 Digging the Postholes. Dig the postholes at the staked locations with a clamshell post-hole digger or a power auger. Center the holes on the string. The holes should be at least 3 feet deep and 10 to 12 inches in diameter.

6 Installing Footings. Footings are the materials placed under and around posts to support them. The footing should always extend below the frost line. Always line the bottom of the hole with a base stone and/or gravel before you put in the post. Gravel prevents concrete from getting under the post where it will seal off the bottom and hold in moisture, decaying the post. Gravel also helps hold the post steady in a vertical position while you tamp more stone or earth around it, or while you pour concrete.

7 Sloping the Concrete Collar. A concrete collar helps rainwater drain away from the post, rather than collecting around it. Pour the concrete to completely fill the hole, and finish it so it slopes away from the post.

8 Bracing the Posts. When using concrete, temporarily brace the posts to keep them steady while the concrete is setting. Several 4-foot lengths of 2x4s, nailed to the post and staked at the ground, should hold the posts firmly. Check the posts for plumb by using a level on two adjacent sides. Let the concrete cure at least 24 hours before adding the rails and fence boards to the posts.

Thrust Down. Open the Handles and Lift Out the Dirt. Break up Any Remaining Rock.

5 Use a clamshell-type post-hole digger as shown above to dig holes at the staked positions.

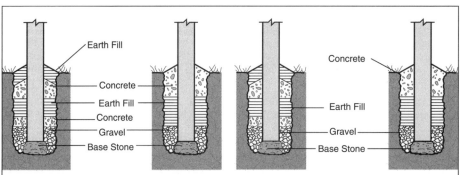

Earth Fill Concrete Earth Fill Concrete Gravel Base Stone Concrete Earth Fill Gravel Base Stone

6 Fence posts must be secured in the ground and protected from groundwater. In most areas, this requires at least some concrete. However, if your soil is firm and not prone to extreme frost heave, you can set a post securely in gravel and earth fill.

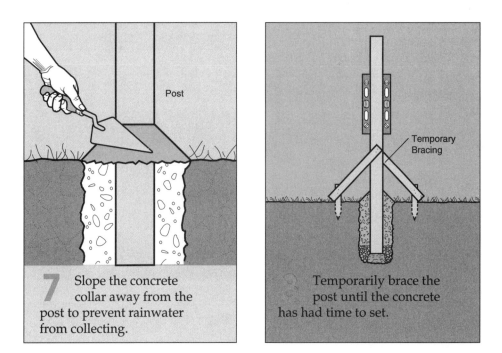

Post

Temporary Bracing

7 Slope the concrete collar away from the post to prevent rainwater from collecting.

8 Temporarily brace the post until the concrete has had time to set.

9 Setting the Posts. As you set each new post in the ground, make sure its height is equal to that of the other posts. To do this, brace the post as just described. Run a string between the posts and use a line level to ensure that the heights are the same. Make any adjustments as needed, then proceed with the pour. Some people prefer to buy longer posts, set them level with fill, then cut off the excess at the top to ensure equal heights.

10 Attaching the Rails. The top rail can be set flush with the top of the posts or dropped several inches below them, depending on the fence style. Nail the top rail inside, outside, or within the frame. If within the frame, it is better to use galvanized hardware such as fence brackets or T-plates rather than simply nailing. Secure the middle (if required) and bottom rails in the same way. For shorter fences, you might want only top and bottom rails.

While many people have a tendency to secure 2x4 rails with the wide side down, this should be avoided. Most of the fence designs can be achieved by laying the rail with the 4-inch side vertical, which provides twice the strength as when laid on its side.

11 Leveling the Rails. For fences that travel over level ground or step down a slope, use a level to make certain that the rails are kept in horizontal alignment.

For fences that follow the slope, secure the top rail flush with the posts, then locate the middle and bottom rails by either measuring down from the top rail when it is in place, or by running a string from post to post to locate the exact position of the rails.

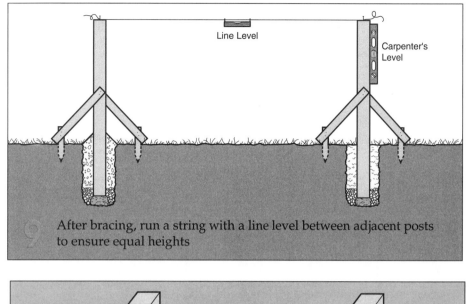

9 After bracing, run a string with a line level between adjacent posts to ensure equal heights

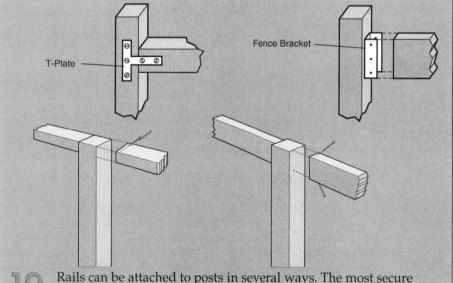

10 Rails can be attached to posts in several ways. The most secure is with galvanized connecting hardware such as T-plates or fence brackets. However, avoid installing rails with the wide side down (bottom left); a much stronger construction utilizes the narrow side down (bottom right).

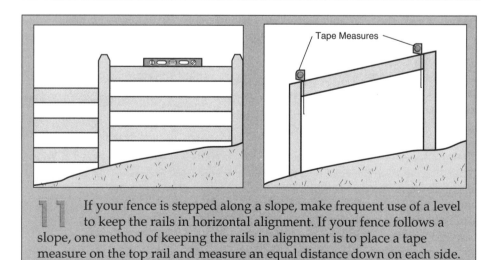

11 If your fence is stepped along a slope, make frequent use of a level to keep the rails in horizontal alignment. If your fence follows a slope, one method of keeping the rails in alignment is to place a tape measure on the top rail and measure an equal distance down on each side.

Working with Concrete

Mixing concrete to make collars for fence posts is not at all difficult. Rather than purchase the cement, sand, and gravel separately, it is much more convenient to buy premixed concrete. One posthole requires about 2/3 cubic foot of wet concrete.

Read the package instructions for the proper ratio of mix to water. Dump the required amount of dry mix in a wheelbarrow and mix it thoroughly with a shovel or hoe. Scoop out a hollow in the middle of the mix and pour in about half of the water. Mix thoroughly, then add the remaining water a little at a time while continuing to stir. Wet concrete should have a stiff consistency; you should be able to pack it into a ball in your hand.

When the concrete is mixed to your satisfaction, pour it into the hole immediately. Compact the concrete with a tamper to displace any air bubbles. Extend the concrete

slightly above soil level, then use a trowel to slope it away from the post for good water drainage. Check the post alignment before the concrete sets, and reposition the post to 90 degrees vertical if necessary. Leave all posts undisturbed for at least 24 hours to allow the concrete to cure fully.

12 **Checking the Plumb.** The installation of the boards on the framework is the easiest part. Do not nail on the boards until the final framework pleases you. When you are ready to nail on the boards, start at either end of the fence and set the first board in place. Be sure it is vertical; use a plumb bob and line to check it.

13 **Adding the Fence Boards.** Board fence material should not overlap the top or bottom rail by more than a few inches. If it does, the unsupported board will warp and cause a ragged-looking edge.

For a privacy fence, butt the boards side by side. If the boards are not being butted against each other, you will need a spacer to ensure even spacing between boards. The spacer can be an actual fence board or just a piece of scrap wood cut to the width the boards will be spaced. (See individual fence projects for specifics.) After you nail up several boards, check with the plumb bob again to be sure you are keeping the boards plumb; the board will not necessarily be perfectly straight, so you will have to adjust as you go along.

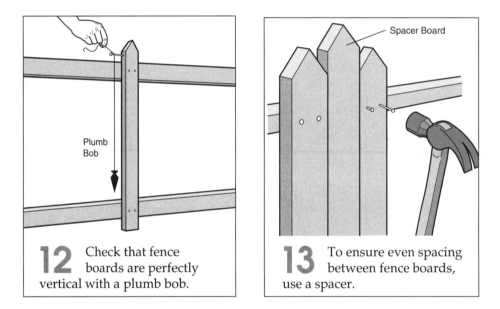

12 Check that fence boards are perfectly vertical with a plumb bob.

13 To ensure even spacing between fence boards, use a spacer.

Turning a Corner with a Fence

Turning a corner is a simple matter. Just treat the corner post as the end post for both of the fence sections; the rails and boards are attached in the standard fashion. Where the top rails meet at the corner post, miter the ends as shown for a neat appearance.

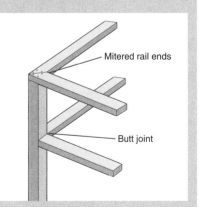

BUILD: FENCES

This section describes the construction of 24 different wood fences. All of these designs are based on the post and rail construction, using a variety of materials and designs for the infill. You can adjust these instructions and materials lists to fit the variations of your own particular design needs.

Picket Fence

Although traditionally associated with Colonial architecture, the picket fence looks good with many different types of homes, including Victorian, Cape Cod, and Ranch-style. While this fence makes an excellent boundary marker, its low height (never more than 4 feet high) and openness allow views both in and out. You may choose to plant shrubs next to the picket fence; ivy, climbing roses, and other vines also work well. This not only provides added privacy, but it lends an attractive touch to the fence as well.

Materials List

3x8 FENCE SECTION

14	1x4x3 pickets
2	2x4x8 rails
2	4x4x6 posts
1/2 lb.	6d galvanized nails
1/4 lb.	16d galvanized nails

The picket fence is easy and fun to build; it requires basic construction skills and minimal labor per section. It is best suited to relatively short runs, such as a housefront; the repetitive design can become monotonous if used in a long, uninterrupted stretch. The picket is often used to enclose a yard, as a garden border (where it can be as little as 2 feet high), and to keep small children from straying. While a picket fence provides little visual privacy, close spacing of the pickets can help soften breezes. Rather than using more expensive treated lumber, a popular choice for the picket fence is pine, which can be painted white for a classic look.

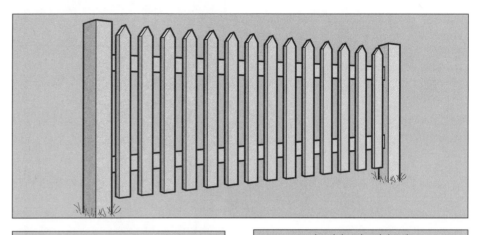

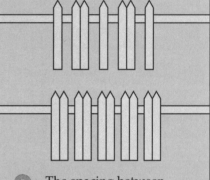

1 Construct the post-and-rail framework as described in steps 1 to 11 of "Build the Basic Fence." Because the picket fence is not high, two rails provide sufficient stability, with the top rail dropped down rather than flush with the post tops.

2 The spacing between pickets is entirely up to you. Most common is a spacing of one picket width. You can also vary the types of picket designs for greater visual effect. (See pages 55 to 57 for more information on picket designs.)

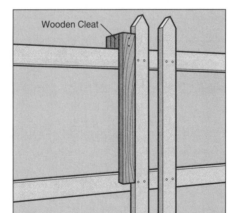

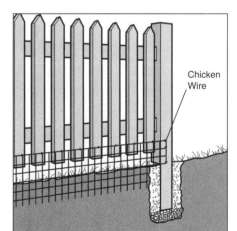

3 Attach each picket with two nails at the top and bottom. A long piece of wood the same width as the pickets can be used for uniform spacing. By nailing a cleat to the top of the piece, it can be hung on the top rail and used as a spacer.

4 If the picket fence is being used to enclose a flower or vegetable garden, you will need added protection against digging animals. An effective technique is to nail chicken wire to the bottom of the fence and bury it a few inches in the soil.

Curved Panel Picket Fence

If there is one drawback to the picket fence, it is that when used in long stretches it can become downright monotonous. These curved pickets solve such a problem by creating a pleasing visual effect. It is fairly easy to cut the curved shape, using a saber saw. While you can also use a keyhole or similar hand saw, doing it this way is more difficult and time-consuming.

Materials List

4x8 FENCE SECTION	
14	1x4x4 pickets
2	2x4x8 rails
2	4x4x7 posts
1/2 lb.	6d galvanized nails
1/4 lb.	16d galvanized nails

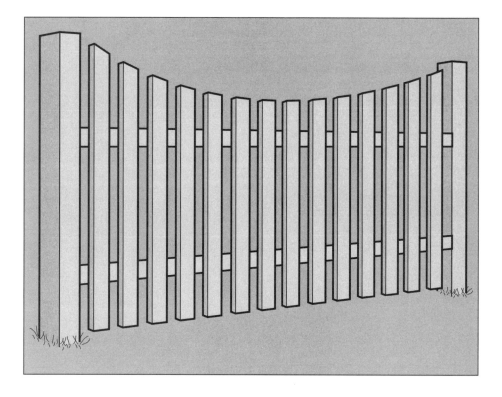

The curved panel picket fence is surprisingly easy to build; it requires basic construction skills and minimal labor per section. Its eye-catching look is best suited to spanning an entire property and other long runs. The most popular uses for the curved panel picket are to enclose a yard, to border a flower garden, and to enclose patios. In addition to its decorative charm, this fence offers a measure of protection and security; it does a nice job of preventing children and pets from wandering.

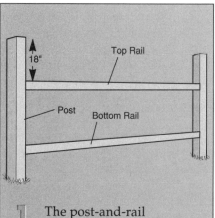

1 The post-and-rail framework for the traditional picket fence has the top rail only a few inches below the post tops. However, the curved panel picket requires the top rail to be approximately 18 inches below the post tops so that it remains hidden when the curve is cut.

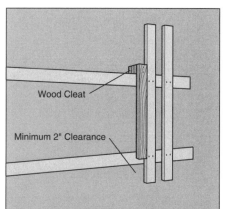

2 Since the tops of the pickets are to be cut, use standard flat tops rather than dog-ears or other decorative tops. Attach the pickets by driving two nails in the top and bottom of each one. Use a piece of wood the same width as the boards for spacing. Nail a cleat to the top of the piece and hang it on the top rail.

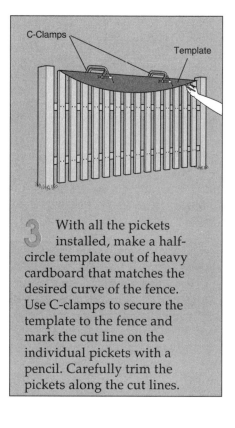

3 With all the pickets installed, make a half-circle template out of heavy cardboard that matches the desired curve of the fence. Use C-clamps to secure the template to the fence and mark the cut line on the individual pickets with a pencil. Carefully trim the pickets along the cut lines.

Louver Fence

The louver fence is a handsome addition to any home's exterior. The pattern of alternating strips of shadow and sunlight change as the sun's angle changes, lending an interesting visual effect. More than any other, this fence is made to be seen, and so should be carefully worked into your landscaping plan. The most common angle for attaching louvers is 45 degrees, but you can set them at any angle you want.

Materials List

4x8 FENCE SECTION
20	1x6x4 boards
2	2x4x8 rails
2	4x4x7 posts
1/2 lb.	8d galvanized nails
1/4 lb.	16d galvanized nails

5x8 FENCE SECTION
20	1x6x5 boards
2	2x4x8 rails
2	4x4x8 posts
1/2 lb.	8d galvanized nails
1/4 lb.	16d galvanized nails

6x8 FENCE SECTION
20	1x6x6 boards
2	2x4x8 rails
2	4x4x9 posts
1/2 lb.	8d galvanized nails
1/4 lb.	16d galvanized nails

The angled boards of the louver fence make it more time-consuming to build than most other fences. It is best suited to short runs; it can be rather expensive to build in long stretches. The louver is best used to enclose a patio, or to border a pool, tennis court, or other recreational facility. It is an excellent choice for hiding unpleasant views. Besides offering visual privacy, a major advantage of a louver fence is its ability to soften and redirect wind currents. This makes it a good choice on oceanfront properties and in other areas susceptible to high winds.

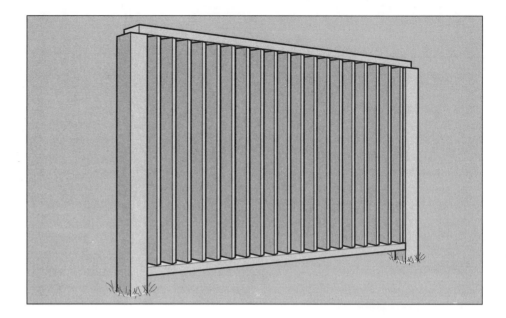

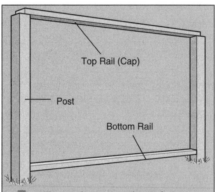

1 Erect the posts (at 8 feet on center) using the procedure described in steps 1 to 9 of "Build the Basic Fence." As with the alternate panel fence, the top rail doubles as the cap. The bottom rail attaches to the inside of the posts, either by toenailing or using fence brackets.

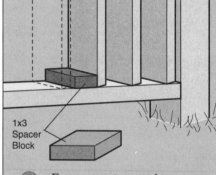

2 For accurate spacing of the boards, make a homemade template by cutting a 1x3 spacer block on the desired angle; a popular choice is 45 degrees. Lay the spacer block on the bottom rail and use it to guide the boards into position.

3 To secure the boards at the top, drive nails through the top rail.

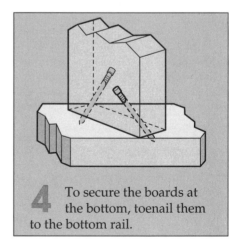

4 To secure the boards at the bottom, toenail them to the bottom rail.

Horizontal Louver Fence

The horizontal louver fence is an interesting variation of the standard louver. Whereas the standard louver uses temporary spacers to accurately position the boards, this fence has permanent cleats fastened to the posts. Also, the horizontal louver uses 1/2-inch siding for the louvers.

Materials List

6x8 FENCE SECTION

16	1/2x6x7' 5" siding louvers
2	2x4x8 rails
2	4x4x9 posts
30	1x3x3 1/2" cleats
1/2 lb.	8d galvanized nails
1/4 lb.	16d galvanized nails

Because it utilizes 30 permanent cleats fastened to the posts, the horizontal louver fence is more labor-intensive than the vertical louver. Like its vertical counterpart, it is best suited to relatively short runs around a patio, or to border a pool, tennis court, or other recreational facility. Horizontal louvers are very sturdy and hard to climb, so they make a good choice when protection and security are foremost requirements.

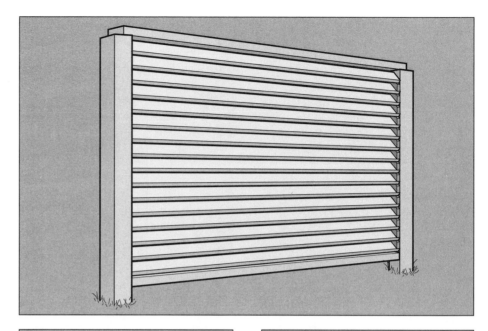

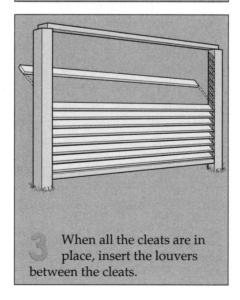

Top Rail (Cap)

Post

Bottom Rail

1 Erect the posts (at 8 feet on center) as described in steps 1 to 9 of "Build the Basic Fence." Add the top rail, or cap, and the bottom rail. Attach the bottom rail either by toenailing or using fence brackets.

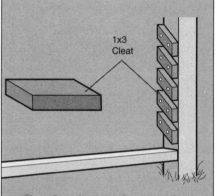

1x3 Cleat

2 Cut the 1x3 cleats on the desired angle; a popular choice is 45 degrees. Starting at the bottom, nail the cleats to the inside of the posts, using two nails per cleat. Leave 1/2 inch in between cleats for the louvers.

3 When all the cleats are in place, insert the louvers between the cleats.

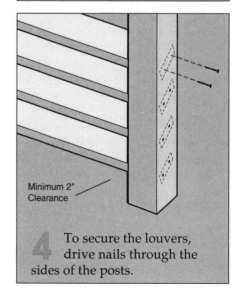

Minimum 2" Clearance

4 To secure the louvers, drive nails through the sides of the posts.

Lattice Top Fence

This eye-catching lattice top fence is easier to build than you might think. Nothing livens up the blankness of a solid wall better than lattice. It is available in preassembled standard widths of 1x8, 2x8, and 4x8. The 1x8 width is perfect for this project.

Materials List

4x8 FENCE SECTION

16	1x6x3 boards
3	2x4x8 rails
2	4x4x7 posts
1	1x8 lattice
8	1x1x8 nailing strips
1/2 lb.	8d galvanized nails
1/4 lb.	16d galvanized nails

5x8 FENCE SECTION

16	1x6x4 boards
3	2x4x8 rails
2	4x4x8 posts
1	1x8 lattice
8	1x1x8 nailing strips
1/2 lb.	8d galvanized nails
1/4 lb.	16d galvanized nails

6x8 FENCE SECTION

16	1x6x5 boards
3	2x4x8 rails
2	4x4x9 posts
1	1x8 lattice
8	1x1x8 nailing strips
1/2 lb.	8d galvanized nails
1/4 lb.	16d galvanized nails

This lattice top fence, though it utilizes preassembled lattice, requires abundant labor per section. It is best suited to relatively short runs due to its delicate form. Lattice is ideal for use in a garden setting; it has long been used in Italian and English gardens. This fence is also ideal for enclosing a backyard, patio, or lanai. Because this version uses lattice in combination with solid boards, it is better suited to provide privacy and security than fences made entirely of lattice.

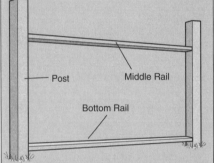

1 Erect the posts (at 8 feet on center) as described in steps 1 to 9 "Build the Basic Fence." Attach the middle and bottom rails inside the posts. Locate the middle rail exactly 1 foot from the top of the posts to allow for the lattice.

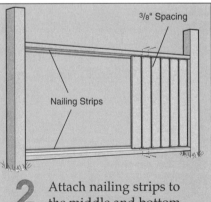

2 Attach nailing strips to the middle and bottom rails. Starting flush against the post, install the boards by toenailing with two nails top and bottom, leaving 3/8-inch between boards.

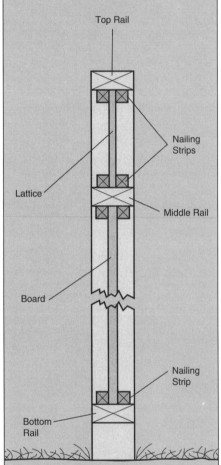

3 Add the second set of nailing strips on the other side of the boards. Next, install the top rail inside the posts. Attach nailing strips, set the lattice in place, then finish with nailing strips on the other side.

Diagonal Board Fence

This sturdy fence with infill panels of diagonally placed boards is as durable as it is attractive. The diagonal pattern helps to lighten the heavy look of the fence. To support the weight of the heavy infill panels, the top and bottom rails are made from 4x6s.

Materials List

6x6 FENCE SECTION

20	1x4x various length diagonals
2	4x6x5' 6 1/2" rails
2	4x6x9 posts
4	2x4x5' 6 1/2" horizontal stops
4	2x4x4' 8" vertical stops
1/2 lb.	6d galvanized nails
1/2 lb.	16d galvanized nails

The diagonal board fence requires advanced construction skills; in fact, it is one of the most challenging fences to build, but the time and effort are well worth it. It is best suited to long runs to help dwarf its bulkiness. Besides its obvious security advantage, the diagonal board fence is particularly well suited to insulating your home or yard from street noise, and functions as an effective barrier to wind. With this fence, you can create an outdoor room or living space for entertaining.

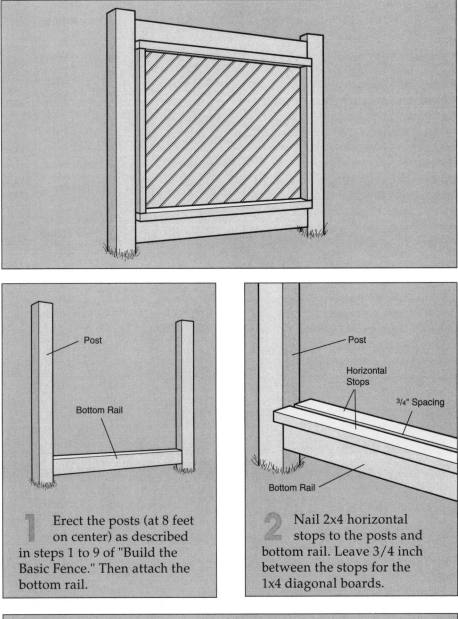

1 Erect the posts (at 8 feet on center) as described in steps 1 to 9 of "Build the Basic Fence." Then attach the bottom rail.

2 Nail 2x4 horizontal stops to the posts and bottom rail. Leave 3/4 inch between the stops for the 1x4 diagonal boards.

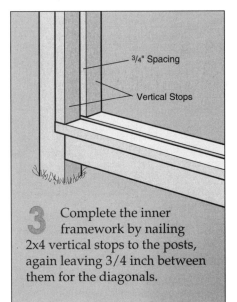

3 Complete the inner framework by nailing 2x4 vertical stops to the posts, again leaving 3/4 inch between them for the diagonals.

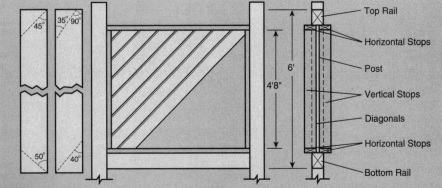

4 Cut the diagonals to size and insert them in the inner framework. Eighteen of the diagonals have identical angled cuts as shown. The two longest diagonals are cut differently at the ends to accommodate their placement in the frame. Secure with nails and wood glue, then attach the top rail to complete the fence section. The diagonals are situated in the inner framework, as illustrated in the side view.

Bamboo Fence

While at first glance the bamboo fence may seem out of place in this section, in actuality it is one of the oldest and most common fences. Bamboo is available at landscape and garden centers. It is a very compatible and attractive material to use in a garden. An interesting effect can be achieved by constructing two separate bamboo fence sections about 3 feet apart and planting a hedge between them. To discourage climbers, you may wish to cut the tops at an angle, with the bottoms of the slants just above the joints.

Bamboo can be tied together with most any type of wire or heavy cord. However, it cannot be nailed, because it splits very easily. While large timber bamboo makes an adequate fence post, sinking it into the ground will cause it to rot fairly quickly. For this reason, it is best to use standard wood posts and keep the bamboo off the ground.

Materials List

4x6 FENCE SECTION

18	1"-dia. x 4 vertical bamboo members
3	3"-dia. x 6 horizontal bamboo members
2	4x4x7 posts
1 spool	wire or heavy cord

The bamboo fence does not use standard fasteners, and requires maximum labor per section due to the extensive tying together of the members. It is a relatively fragile fence that is best suited to short runs. Used in combination with landscaping shrubs and plants, a bamboo fence is ideal for the outdoor gardener. For privacy, just lash the vertical members close together instead of spacing them. The bamboo fence requires very little maintenance, just occasional inspection for rotting members and insect infestation.

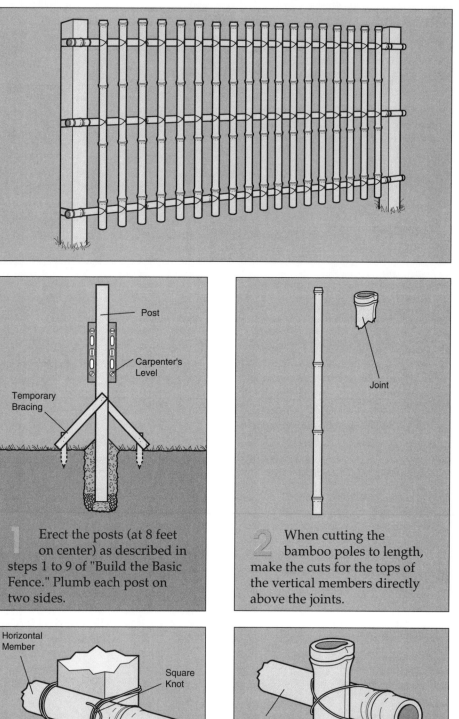

1 Erect the posts (at 8 feet on center) as described in steps 1 to 9 of "Build the Basic Fence." Plumb each post on two sides.

2 When cutting the bamboo poles to length, make the cuts for the tops of the vertical members directly above the joints.

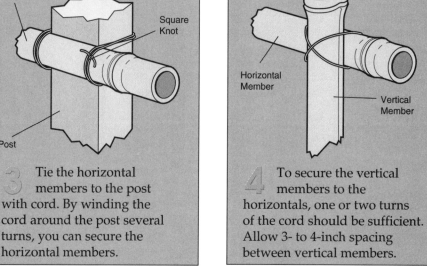

3 Tie the horizontal members to the post with cord. By winding the cord around the post several turns, you can secure the horizontal members.

4 To secure the vertical members to the horizontals, one or two turns of the cord should be sufficient. Allow 3- to 4-inch spacing between vertical members.

Alternate Width Fence

Alternating 1x4 and 1x6 boards are used to create this attractive fence that looks good from either side. In addition to the alternating widths, the 1x4s are cut shorter to provide an added dimension of alternating heights. This fence lets in a fair amount of sunshine while still offering partial wind resistance.

Materials List

3'4"x8 FENCE SECTION	
16	1x6x3' 4" boards
15	1x4x3 boards
2	2x4x8 rails
2	4x4x6 posts
1/2 lb.	8d galvanized nails
1/4 lb.	16d galvanized nails

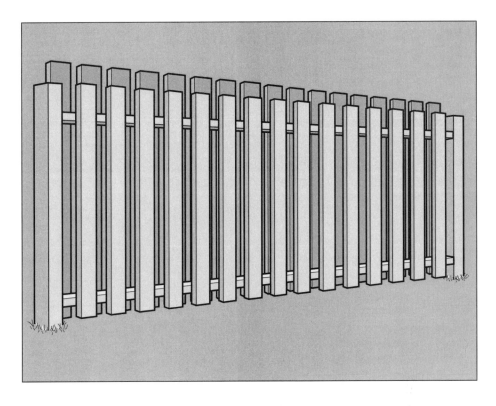

The alternate width fence requires basic construction skills. It is best suited to long runs, such as bordering a long yard or field, in order to achieve the visual effect intended by the varying board lengths. The alternate width fence is best used for demarking a boundary or property line. It is also effective *for providing a measure of privacy. Shrubs and other plantings may be desired to soften the hard look of the wood.*

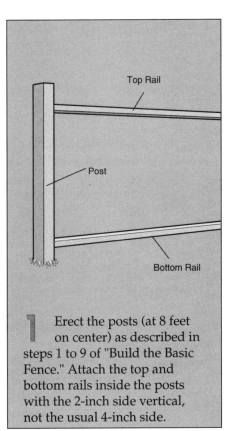

1 Erect the posts (at 8 feet on center) as described in steps 1 to 9 of "Build the Basic Fence." Attach the top and bottom rails inside the posts with the 2-inch side vertical, not the usual 4-inch side.

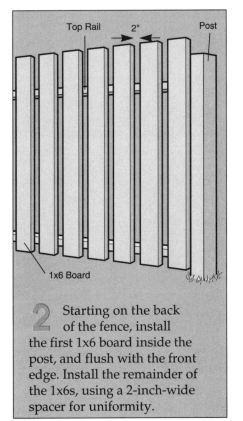

2 Starting on the back of the fence, install the first 1x6 board inside the post, and flush with the front edge. Install the remainder of the 1x6s, using a 2-inch-wide spacer for uniformity.

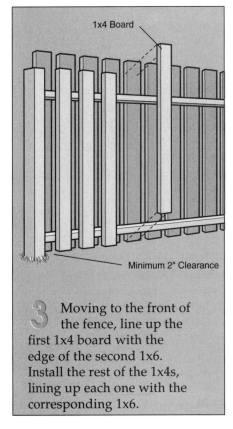

3 Moving to the front of the fence, line up the first 1x4 board with the edge of the second 1x6. Install the rest of the 1x4s, lining up each one with the corresponding 1x6.

Basketweave Fence

The basketweave is mainly used for short lengths of fence. It looks good from both sides, making it an appealing privacy fence. The interwoven design is surprisingly strong for its light weight.

There are many variations of the basketweave design, but none use conventional rails. The design can be horizontal or vertical, and the weave can vary from flat to very wide and open. One popular version uses very thin, lightweight strips, crisscrossed in a tight weave.

Materials List

4x8 FENCE SECTION

9	1x6x8 boards
1	2x4x4 support
1	2x4x8 cap
2	4x4x7 posts
2	2x2x4 nailers
1/2 lb.	8d galvanized nails
1/4 lb.	16d galvanized nails

5x8 FENCE SECTION

11	1x6x8 boards
1	2x4x5 support
1	2x4x8 cap
2	4x4x8 posts
2	2x2x5 nailers
1/2 lb.	8d galvanized nails
1/4 lb.	16d galvanized nails

6x8 FENCE SECTION

13	1x6x8 boards
1	2x4x6 support
1	2x4x8 cap
2	4x4x9 posts
2	2x2x6 nailers
1/2 lb.	8d galvanized nails
1/4 lb.	16d galvanized nails

The basketweave fence requires a fair amount of skill and patience to construct. It is best suited to short runs, such as between your yard and your neighbor's yard, though it can be a bit overwhelming if used to border a very small area. This fence is often used for demarking a boundary or property line, and comes in handy for hiding unpleasant views. It makes an excellent fence for privacy.

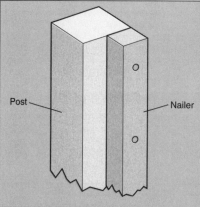

1 Erect the posts (at 8 feet on center) as described in steps 1 to 9 of "Build the Basic Fence." Affix the 2x2 nailers to the inside of each post.

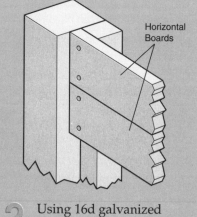

2 Using 16d galvanized common nails, secure the horizontal boards to the nailers just attached. Be sure to butt the edges of the boards against each other.

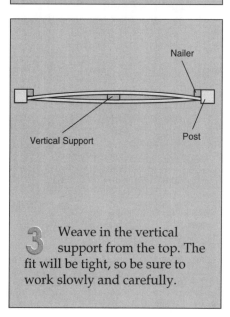

3 Weave in the vertical support from the top. The fit will be tight, so be sure to work slowly and carefully.

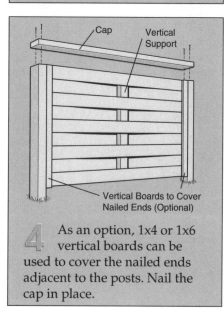

4 As an option, 1x4 or 1x6 vertical boards can be used to cover the nailed ends adjacent to the posts. Nail the cap in place.

Jasper Fence

The jasper fence is similar to the picket fence; its clean lines and openness make it a popular choice with homeowners. However, while the picket only reaches 4 feet high, a jasper fence is often built as high as 6 feet.

Materials List

4x8 FENCE SECTION

16	1x6x4 boards
2	2x4x8 rails
2	4x4x7 posts
1/2 lb.	8d galvanized nails
1/4 lb.	16d galvanized nails

5x8 FENCE SECTION

16	1x6x5 boards
2	2x4x8 rails
2	4x4x8 posts
1/2 lb.	8d galvanized nails
1/4 lb.	16d galvanized nails

6x8 FENCE SECTION

16	1x6x6 boards
2	2x4x8 rails
2	4x4x9 posts
1/2 lb.	8d galvanized nails
1/4 lb.	16d galvanized nails

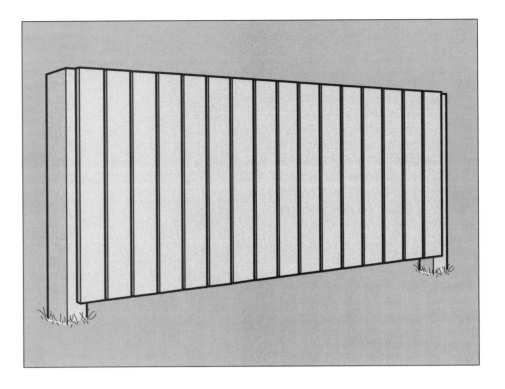

The jasper fence is quite easy to build. It is best suited for use as a border for patios, swimming pools, and other areas around the home where privacy is desired. The simplicity of the jasper makes it very adaptable, and thus equally suited to both long and short runs. Because of its solid-board sections, it is one of the best choices you can make for both security and noise control.

1 Construct a post-and-rail framework identical to that used for the picket fence.

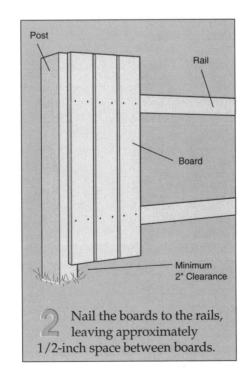

2 Nail the boards to the rails, leaving approximately 1/2-inch space between boards.

Post

Rail

Board

Minimum 2" Clearance

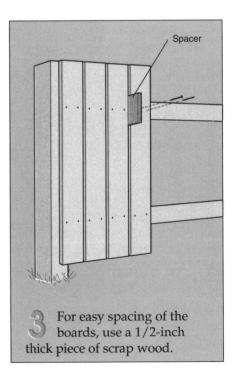

3 For easy spacing of the boards, use a 1/2-inch thick piece of scrap wood.

Spacer

Horizontal Board & Board Fence

The horizontal board and board fence offers many of the advantages of the louver fence, but it is easier and less costly to build. The baffle-like arrangement of the boards helps break up strong wind currents, but still allows air to circulate freely. The staggered construction creates interesting shadow patterns as the sun advances throughout the course of the day, in contrast to straight board fences.

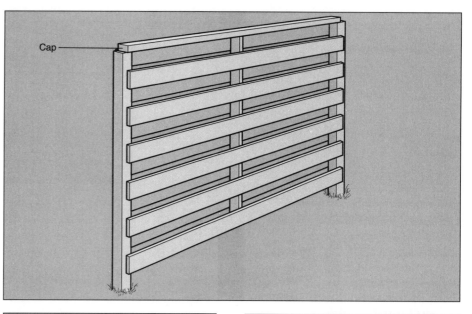

Materials List

4x8 FENCE SECTION
8	1x6x8 boards
1	2x4x4 support
1	2x4x8 cap
2	4x4x7 posts
1/2 lb.	8d galvanized nails
1/4 lb.	16d galvanized nails

5x8 FENCE SECTION
10	1x6x8 boards
1	2x4x5 support
1	2x4x8 cap
2	4x4x8 posts
1/2 lb.	8d galvanized nails
1/4 lb.	16d galvanized nails

6x8 FENCE SECTION
12	1x6x8 boards
1	2x4x6 support
1	2x4x8 cap
2	4x4x9 posts
1/2 lb.	8d galvanized nails
1/4 lb.	16d galvanized nails

The horizontal board and board fence requires a moderate amount of labor per section. It is best suited to short runs, such as bordering a garden, patio, or veranda. Because of its open-board sections, the horizontal board and board is not the best choice for a wind barrier, but it does offer a measure of privacy because the boards are staggered.

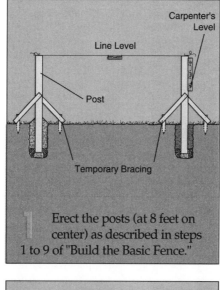

1 Erect the posts (at 8 feet on center) as described in steps 1 to 9 of "Build the Basic Fence."

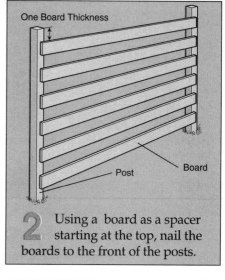

2 Using a board as a spacer starting at the top, nail the boards to the front of the posts.

3 Nail the center support to the back of the boards installed. On the opposite side, nail the boards to the back of the posts and the center support.

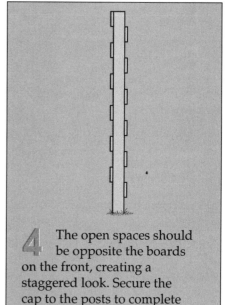

4 The open spaces should be opposite the boards on the front, creating a staggered look. Secure the cap to the posts to complete the fence section.

Vertical Board & Board Fence

This vertical board and board is an attractive privacy fence that utilizes two 2x4x8 rails and sixteen 1x6 boards, in addition to the basic posts. Known in some circles as a "good neighbor" fence, the vertical board and board provides an effective boundary marker or divider without totally blocking your view.

Materials List

4x8 FENCE SECTION

16	1x6x4 boards
2	2x4x8 rails
2	4x4x7 posts
1/2 lb.	8d galvanized nails
1/4 lb.	16d galvanized nails

5x8 FENCE SECTION

16	1x6x5 boards
2	2x4x8 rails
2	4x4x8 posts
1/2 lb.	8d galvanized nails
1/4 lb.	16d galvanized nails

6x8 FENCE SECTION

16	1x6x6 boards
2	2x4x8 rails
2	4x4x9 posts
1/2 lb.	8d galvanized nails
1/4 lb.	16d galvanized nails

Like its horizontal counterpart, the vertical board and board fence requires basic construction skills. It is ideal for bordering gardens, patios, verandas, and similar short run applications. Because of its open-board sections, the vertical board and board is a poor wind barrier, but its staggered board construction does provide moderate privacy.

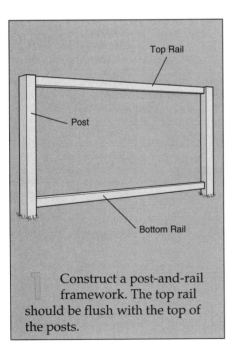

Top Rail

Post

Bottom Rail

1 Construct a post-and-rail framework. The top rail should be flush with the top of the posts.

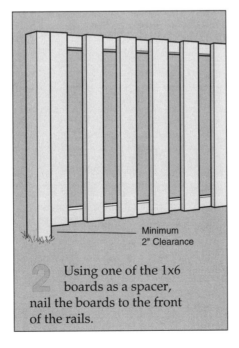

Minimum 2" Clearance

2 Using one of the 1x6 boards as a spacer, nail the boards to the front of the rails.

3 Nail the boards to the back of the rails, opposite the boards on the front.

Solid Board Fence

This solid board fence has an almost formal look. It functions very well to define work and play areas. No fence does a better job of providing maximum privacy, and plantings can be used to soften the imposing look of the wood.

Materials List

4x8 FENCE SECTION
9	1x6x8 boards
1	2x4x4 support
2	2x4x8 rails
2	4x4x7 posts
2	2x2x4 nailers
1/2 lb.	8d galvanized nails
1/4 lb.	16d galvanized nails

5x8 FENCE SECTION
11	1x6x8 boards
1	2x4x5 support
2	2x4x8 rails
2	4x4x8 posts
2	2x2x5 nailers
1/2 lb.	8d galvanized nails
1/4 lb.	16d galvanized nails

6x8 FENCE SECTION
13	1x6x8 boards
1	2x4x6 support
2	2x4x8 rails
2	4x4x9 posts
2	2x2x6 nailers
1/2 lb.	8d galvanized nails
1/4 lb.	16d galvanized nails

The solid board fence requires moderate labor per section, and is unsurpassed in providing security and privacy. It is best suited to short to medium runs, since its closed construction can be a bit stifling when used in particularly long stretches. Because of its solid-board sections, it is a good wind and noise barrier, and can be used to keep small children and pets from straying.

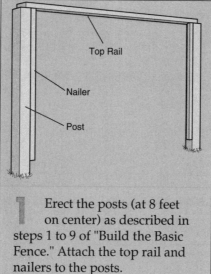

1 Erect the posts (at 8 feet on center) as described in steps 1 to 9 of "Build the Basic Fence." Attach the top rail and nailers to the posts.

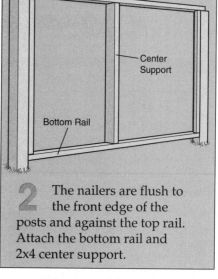

2 The nailers are flush to the front edge of the posts and against the top rail. Attach the bottom rail and 2x4 center support.

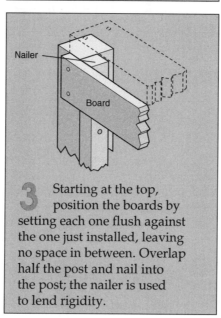

3 Starting at the top, position the boards by setting each one flush against the one just installed, leaving no space in between. Overlap half the post and nail into the post; the nailer is used to lend rigidity.

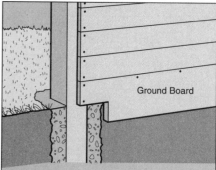

4 To keep children or pets in (or, in the case of a garden, destructive animals out), install a 1x8 or 1x10 treated ground board along the fence bottom, several inches below ground.

Dog-Ear Privacy Fence

To put it simply, the dog-ear fence is a taller version of a picket fence without the spaces. It is easy to build, provides all the privacy you want, and goes well with almost any style house.

Materials List

4x8 FENCE SECTION

17	1x6x4 boards
3	2x4x8 rails
2	4x4x7 posts
1/2 lb.	8d galvanized nails
1/4 lb.	16d galvanized nails

5x8 FENCE SECTION

17	1x6x5 boards
3	2x4x8 rails
2	4x4x8 posts
1/2 lb.	8d galvanized nails
1/4 lb.	16d galvanized nails

6x8 FENCE SECTION

17	1x6x6 boards
3	2x4x8 rails
2	4x4x9 posts
1/2 lb.	8d galvanized nails
1/4 lb.	16d galvanized nails

The dog-ear privacy fence requires basic construction skills and moderate labor per section. Because privacy and security are the sole reasons for erecting this fence, it is most often used in long runs. It is sometimes built higher than 6 feet in order to make a more imposing and effective barrier. The dog-ear privacy fence is also an effective property marker and border fence.

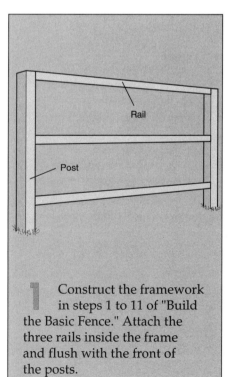

1 Construct the framework in steps 1 to 11 of "Build the Basic Fence." Attach the three rails inside the frame and flush with the front of the posts.

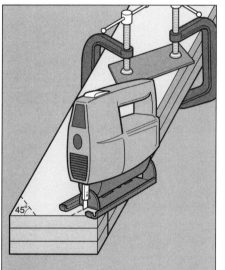

2 To quickly cut dog-ear tops with a saber saw, clamp several boards together, with the top board marked for cutting. Save the top board as a master for cutting additional tops.

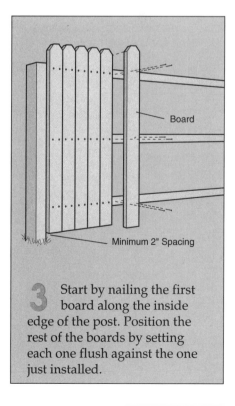

3 Start by nailing the first board along the inside edge of the post. Position the rest of the boards by setting each one flush against the one just installed.

Alternate Panel Fence

Here is a same-on-both-sides fence that is great for privacy and security. It is a popular choice for patios and backyards where you want to create your own private space for reading, sunbathing, etc. And you can use the bottom "lip" for holding plants.

Materials List

4x8 FENCE SECTION

17	1x6x4 boards
1	2x4x4 support
2	2x4x8 rails
2	4x4x7 posts
1/2 lb.	8d galvanized nails
1/4 lb.	16d galvanized nails

5x8 FENCE SECTION

17	1x6x5 boards
1	2x4x5 support
2	2x4x8 rails
2	4x4x8 posts
1/2 lb.	8d galvanized nails
1/4 lb.	16d galvanized nails

6x8 FENCE SECTION

17	1x6x6 boards
1	2x4x6 support
2	2x4x8 rails
2	4x4x9 posts
1/2 lb.	8d galvanized nails
1/4 lb.	16d galvanized nails

In addition to being one of the sturdiest fences you can build, the alternate panel lets you share an attractive fence with your neighbors. This fence style goes well with contemporary architecture. The alternate panel is best suited to short runs; many homeowners use just one or two sections as a garden or patio screen. Yet another solid-board fence, the alternate panel is a good choice for wind and noise control.

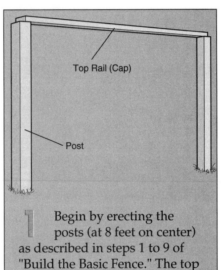

1 Begin by erecting the posts (at 8 feet on center) as described in steps 1 to 9 of "Build the Basic Fence." The top rail doubles as the cap.

Top Rail (Cap)

Post

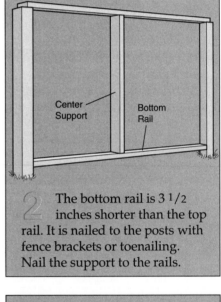

2 The bottom rail is 3 1/2 inches shorter than the top rail. It is nailed to the posts with fence brackets or toenailing. Nail the support to the rails.

Center Support

Bottom Rail

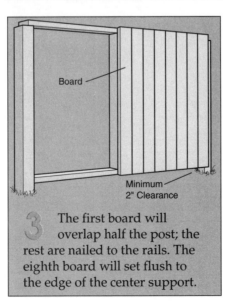

3 The first board will overlap half the post; the rest are nailed to the rails. The eighth board will set flush to the edge of the center support.

Board

Minimum 2" Clearance

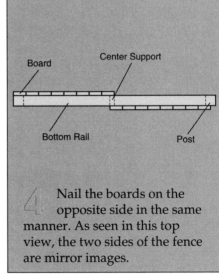

4 Nail the boards on the opposite side in the same manner. As seen in this top view, the two sides of the fence are mirror images.

Board

Center Support

Bottom Rail

Post

Grapestake Fence

Made from redwood logs, grapestakes are about 2 inches square and can range in length from 3 to 6 feet. Their irregular, splintery edges give them a truly Western look, and since redwood offers superior resistance to decay, little maintenance is required. Grapestakes weather to a soft gray that blends well with plantings, as well as brick and stonework. Their light weight makes them easy to handle and to install.

Materials List

4x8 FENCE SECTION

48	2x2x4 grapestakes
2	2x4x8 rails
2	4x4x7 posts
3/4 lb.	6d galvanized nails

5x8 FENCE SECTION

48	2x2x5 grapestakes
2	2x4x8 rails
2	4x4x8 posts
3/4 lb.	6d galvanized nails

6x8 FENCE SECTION

48	2x2x6 grapestakes
2	2x4x8 rails
2	4x4x9 posts
3/4 lb.	6d galvanized nails

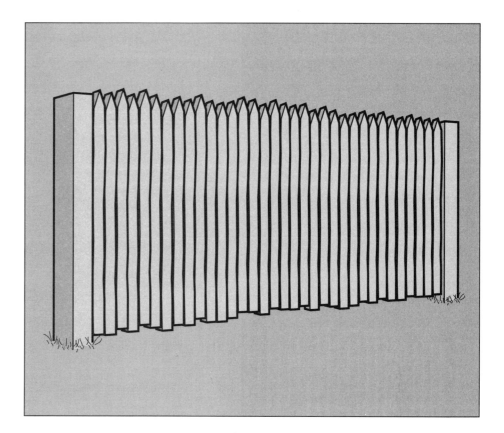

The grapestake fence can be built as easily and almost as quickly as a picket. It is most attractive when viewed from a distance, and therefore is best suited to relatively long runs. The grapestake blocks noise and provides shade, and is a surprisingly strong fence. This design lends itself to the garden alone or to surround an entire acre around your house.

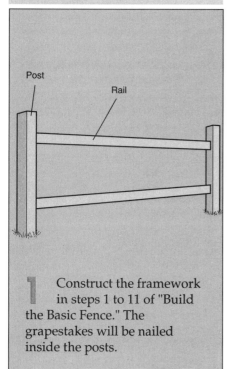

1 Construct the framework in steps 1 to 11 of "Build the Basic Fence." The grapestakes will be nailed inside the posts.

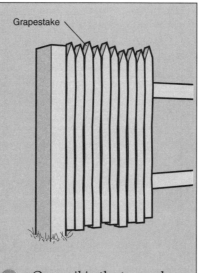

2 One nail in the top and bottom of each grapestake will secure it to the rails. Butt the grapestakes flush against each other, with the pointed tips facing up.

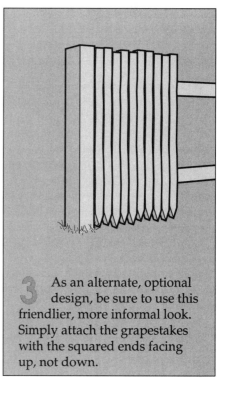

3 As an alternate, optional design, be sure to use this friendlier, more informal look. Simply attach the grapestakes with the squared ends facing up, not down.

Closed Post & Board Fence

The closed post and board fence is ideal for keeping children and pets in without hiding a pleasant view. Easy to build, it uses 1x3 slats as a substitute for rails. These slats let in sunlight and breezes, making the closed post and board the perfect backyard fence.

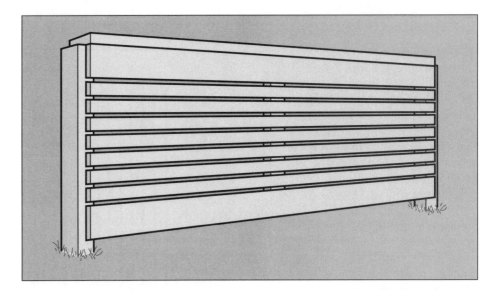

Materials List

4x8 FENCE SECTION
2	1x6x8 boards
10	1x3x8 boards
1	2x4x4 support
1	2x4x8 cap
2	4x4x7 posts
3/4 lb.	8d galvanized nails

5x8 FENCE SECTION
2	1x6x8 boards
14	1x3x8 boards
1	2x4x5 support
1	2x4x8 cap
2	4x4x8 posts
3/4 lb.	8d galvanized nails

6x8 FENCE SECTION
2	1x6x8 boards
17	1x3x8 boards
1	2x4x6 support
1	2x4x8 cap
2	4x4x9 posts
3/4 lb.	8d galvanized nails

The closed post and board fence requires moderate labor per section. It is best suited to short runs when used around the home, but in open country settings it works well in longer stretches. It is a good choice to enclose a patio or small side yard. While the closed post and board gives the appearance of security, in reality it is an easy fence to climb over.

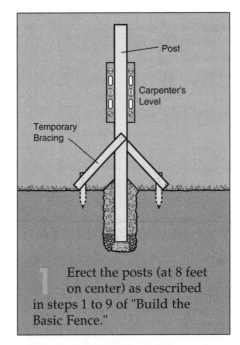

1 Erect the posts (at 8 feet on center) as described in steps 1 to 9 of "Build the Basic Fence."

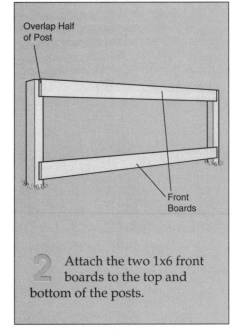

2 Attach the two 1x6 front boards to the top and bottom of the posts.

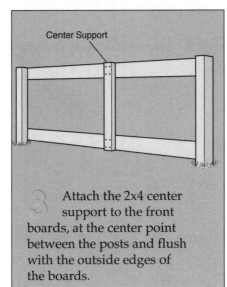

3 Attach the 2x4 center support to the front boards, at the center point between the posts and flush with the outside edges of the boards.

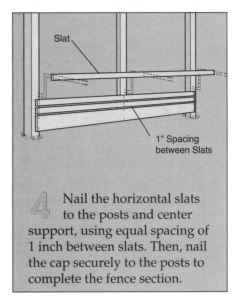

4 Nail the horizontal slats to the posts and center support, using equal spacing of 1 inch between slats. Then, nail the cap securely to the posts to complete the fence section.

Plywood Siding Fence

Exterior-grade plywood is ideal for use as a security fencing material. It comes a standard width of 4 feet, and lengths of 6, 8, and 10 feet. Exterior-grade plywood is available rough sawn, grooved, channeled, and lapped; in addition, it may be unfinished, primed for painting, or prestained in the color of your choice. The fence section shown uses a larger piece of vertically grooved plywood sheet cut down to 4x6, with the grooves now on the diagonal. By grooving the posts, the plywood siding fence assembles quickly and easily.

Materials List

6x4 FENCE SECTION

1	5/8x4x6 plywood sheet
2	4x4x10 posts
2	2x2x4 nailers
1/2 lb.	10d galvanized nails

The plywood siding fence requires basic construction skills, but be sure to choose a material thick enough to resist bowing in heavy wind. This fence can be overbearing in confined spaces, and so is best suited to long runs. Its solid panels afford privacy without the heavy, bulky fence sections previously discussed. Use the plywood siding fence to line your property or surround a swimming pool or other recreational area.

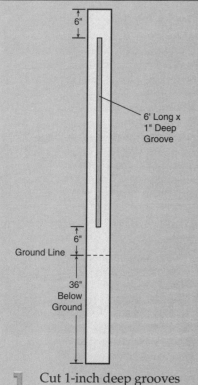

6"

6' Long x 1" Deep Groove

6"

Ground Line

36" Below Ground

1 Cut 1-inch deep grooves with a router, a circular saw, or chisel. If more than one fence section is being made, each post must be routed twice, on opposite sides. The end post needs only one groove.

Plywood Panel

Minimum 2" Clearance

2 Erect the posts, as in steps 1 to 9 of "Build the Basic Fence." Insert one side of the plywood sheet. With the aid of a helper, bow the sheet and pry the other side into the opposite groove.

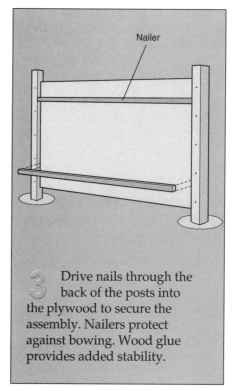

Nailer

3 Drive nails through the back of the posts into the plywood to secure the assembly. Nailers protect against bowing. Wood glue provides added stability.

Alternate Horizontal & Vertical Fence

This truly interesting design is essentially a combination of the solid board fence, with its horizontal boards, and the alternate panel fence, with its vertical boards. For this reason, one section of the alternate horizontal and vertical fence is in actuality two sections, and thus a full 16 feet wide.

Materials List

6x16 FENCE SECTION

13	1x6x8 horizontal boards
16	1x6x6 vertical boards
5	2x4x8 rails
1	2x4x6 support
2	2x2x6 nailers
3	4x4x9 posts
1 lb.	8d galvanized nails
1/2 lb.	16d galvanized nails

The alternate horizontal and vertical fence requires a considerable amount of labor per section, considering that a section is twice as long as those of other fences. It is best suited to long runs, since this helps to achieve the full visual effect of the pattern. Use the alternate horizontal and vertical fence to enclose several acres or a backyard that is a considerable expanse. It also provides moderate security and privacy.

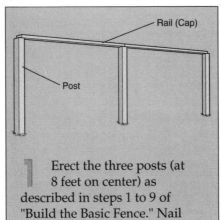

1 Erect the three posts (at 8 feet on center) as described in steps 1 to 9 of "Build the Basic Fence." Nail two of the 2x4 rails to the top of the posts as caps.

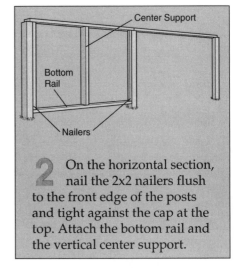

2 On the horizontal section, nail the 2x2 nailers flush to the front edge of the posts and tight against the cap at the top. Attach the bottom rail and the vertical center support.

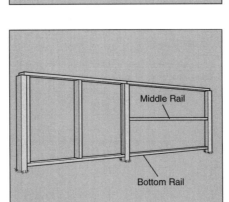

3 Moving to the vertical section, attach the middle and bottom rails inside the posts, either by toenailing or using fence brackets. Nailers are not required with this section. Begin to nail the horizontals. Start at the top and butt the horizontal boards one against the other.

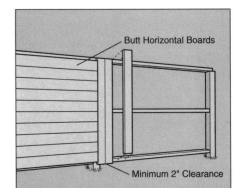

4 Begin the verticals by overlapping half the center post and nailing into the post and nailer. Nail the 16 vertical boards in place. Unlike the alternate panel fence, all of the boards are nailed to the front of the fence. The first and last board will overlap the posts; the rest are nailed into the rails.

Ranch Rail Fence

The rustic attractiveness of this ranch rail fence makes it an attractive boundary marker for large properties. If you wish, 6-inch-diameter round posts can be used instead of the typical 4x4 square posts.

If appearance is your main concern, notch the posts as described in the following steps. If you are more concerned with utility than appearance, you can attach the rails by simply lapping them over the posts and nailing them in place. The completed ranch rail fence section shown here is the most elaborate version: the posts are notched, three rails are included, and a cap tops off the fence.

Materials List

4x8 FENCE SECTION

3	1x6x8 boards
1	1x4x8 boards
2	4x4x7 posts
1/2 lb.	6d galvanized nails

The ranch rail fence requires advanced construction methods, but only minimal labor per section. It is best suited to long runs, such as enclosing several acres around your home or the perimeter of your entire property. It offers no visual privacy, and is too inviting to climbers. Most commonly used in rural applications, the ranch rail is ideal for livestock or setting off a meadow or pasture.

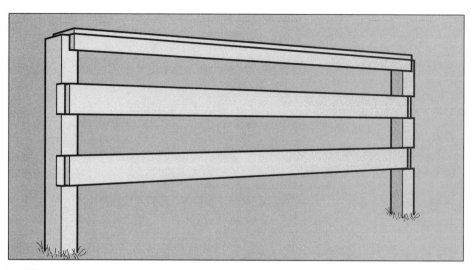

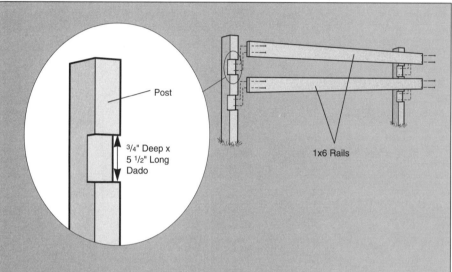

1. Before the posts are erected, cut a 3/4-inch deep x 5 1/2-inch long dado in each post with a router to accommodate the 1x6 rails. See "Cutting a Dado," page 21. A double fluted bit is best for routing this size dado. Then erect the posts as described in steps 1 to 9 of "Build the Basic Fence." Nail the 1x6 rails into the post dados.

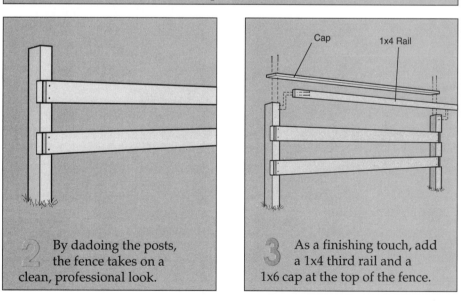

2. By dadoing the posts, the fence takes on a clean, professional look.

3. As a finishing touch, add a 1x4 third rail and a 1x6 cap at the top of the fence.

Post & Rail Fence

The post and rail fence is an interesting variation of the ranch rail design. Its old-fashioned look is still popular in many areas for enclosing fields, orchards, and livestock pens; it also provides a decorative border for ranch-style homes. The post and rail is a very sturdy and durable fence, making it an ideal choice for homeowners.

Materials List

4x8 FENCE SECTION

2	1x6x10 boards
2	1x6x8 boards
2	4x4x7 posts
1/2 lb.	6d galvanized nails

The post and rail fence requires basic construction skills. Like the ranch rail, it is best suited to long runs, such as enclosing several acres around your home or the perimeter of your entire property. As is the case with other rail fences, the post and rail offers no visual privacy or protection from the elements, with the exception of its ability to block drifting snow.

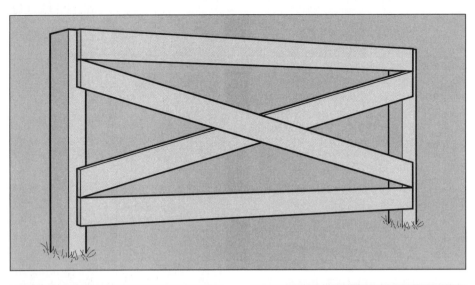

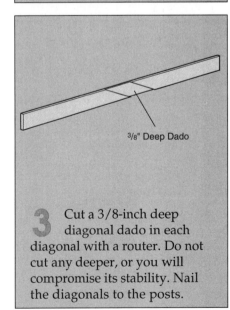

1 Erect the posts (at 8 feet on center) as described in steps 1 to 9 of "Build the Basic Fence." Be sure to use temporary bracing on the posts until the concrete has set.

Diagonals

2 Next, the diagonals must be dadoed. Position the diagonals as they will be in the fence, clamp securely, and mark the overlapping portions to be dadoed.

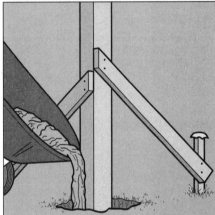

3/8" Deep Dado

3 Cut a 3/8-inch deep diagonal dado in each diagonal with a router. Do not cut any deeper, or you will compromise its stability. Nail the diagonals to the posts.

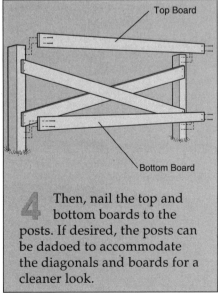

Top Board

Bottom Board

4 Then, nail the top and bottom boards to the posts. If desired, the posts can be dadoed to accommodate the diagonals and boards for a cleaner look.

Split Rail Fence

The split rail fence has long been used for cattle and horse pastures, but this does not mean it should be confined only to rural use. It also makes a popular decorative fence, and goes well with Western-style homes. The split rail fence utilizes 6-inch diameter logs for the posts, and 4-inch diameter logs for the rails. While the design shown here has three rails, the fence works just as well with only the top two.

Materials List

4x8 FENCE SECTION	
3	4"-diameter x 8 rails
2	6"-diameter x 7 posts
1/2 lb.	16d galvanized nails

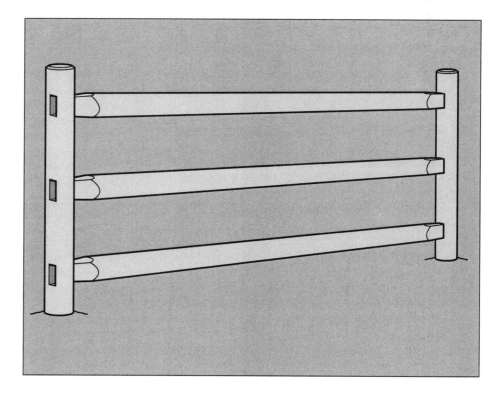

The split rail fence requires rather advanced construction methods, and as such can be a time-consuming fence to erect. It is used almost exclusively in long runs, where its bold and broad outline creates an interesting perimeter boundary. Use the split rail to enclose several acres around your home or your entire property It is also a beautiful choice for use along a country road.

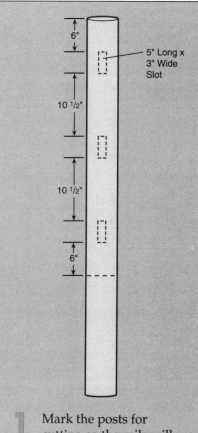

1 Mark the posts for cutting so the rails will be evenly spaced, 10 1/2 inches apart. Use a drill and chisel to mortise three slots in each post. Refer to page 50 for complete mortising details.

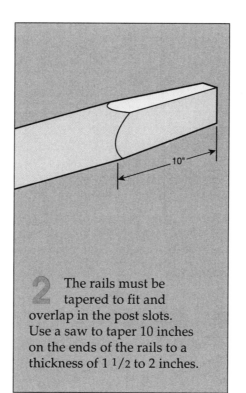

2 The rails must be tapered to fit and overlap in the post slots. Use a saw to taper 10 inches on the ends of the rails to a thickness of 1 1/2 to 2 inches.

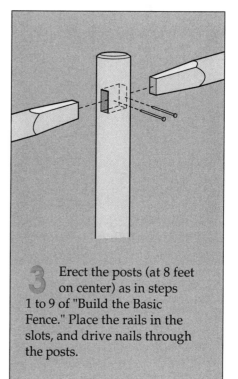

3 Erect the posts (at 8 feet on center) as in steps 1 to 9 of "Build the Basic Fence." Place the rails in the slots, and drive nails through the posts.

This procedure takes some practice to master. Try it first on smaller scrap wood before moving on to the fence.

1. Lay out the tenon by measuring from the end of the rail the length of the tenon; in the case of the "Mortised Rail Fence" the tenon should be half the thickness of the post, or 1 3/4 inches. Use a try square to mark this line on all four sides of the rail. The tenon should be one third to one half the thickness of the piece on which it is being cut. Since the 2x6 is actually 1 1/2 inches thick, make the tenon at least 3/4 inch thick. Locate the exact center along one edge and measure half of the tenon thickness on each side of this center point. Use a try square to extend these markings the length of the tenon, across the end, and down the other side to the previously marked shoulder lines indicating the end of the tenon.

2. Follow the same steps to mark the width of the tenon on the post (usually a cut of at least 3/4 inch at the top and 1/4 inch at the bottom). Use a dovetail saw or a back saw to cut along these guidelines to the depth of the shoulder lines. Cut the back line and part of the top first, then turn the piece around and cut the rest of the top and the front lines.

3. The length of the mortise must equal the width of the tenon. With the try square and a pencil, mark lines across the post to indicate mortise length. The joint is strongest when the cheek thicknesses added together equal the mortise width. Locate the exact center of the mortise and measure out from the center precisely half the width of the mortise (which corresponds to the thickness of the tenon). Outline the mortise, then check your markings by placing the tenon on the surface of the post. Score the outline with a sharp chisel.

4. Choose a drill bit slightly smaller than the width of the mortise. Place the spur of the bit on the center line of the mortise and bore a series of overlapping holes, with the first hole barely touching one end of the mortise and the last hole barely touching the other end. Hold the bit perfectly perpendicular to the work during this operation. For a through mortise-and-tenon joint, the holes should be bored completely through the wood. Then use a small chisel and hammer to square off the corners of the mortise.

5. With a sharp chisel, clean the walls of the mortise to the guidelines, keeping them perpendicular to the surface of the work. To clean out the mortise cavity, strike it hard with the chisel (left), then lever out the waste (right).

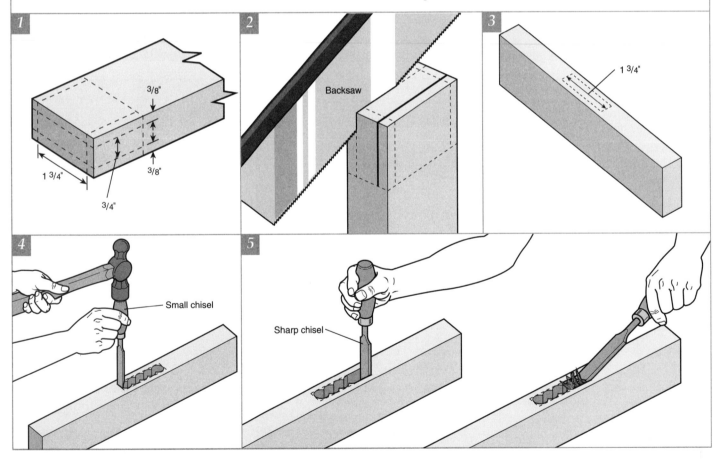

Mortised Rail Fence

The mortised rail fence is a close relative of the split rail fence. At first glance, the only difference appears to be the fact that the mortised rail is built with standard lumber instead of the more rustic logs used in the split rail. However, the mortise-and-tenon joint is a much more involved method of fastening. This joint can be a bit challenging for the first-timer; its strength depends on the accurate fitting of tenon to mortise. Once mastered, however, it produces a clean, neat, and exceptionally strong joint.

A mortise is a rectangular cavity cut into a piece of wood, into which a tenon (a projection cut on the end of a second piece of wood) is inserted. A through mortise, which is used in the mortised rail fence, extends all the way through the wood. The tenon for a through mortise is normally cut slightly longer than the thickness of the mortised piece and then trimmed off flush after assembly, but in this case it is cut to only half the thickness; this will allow for the rails in the adjacent fence section.

Materials List

4x8 FENCE SECTION	
3	2x6x8 rails
2	4x4x7 posts
1/2 lb.	16d galvanized nails

Like the split rail fence, the mortised rail requires advanced construction methods and abundant labor per section. It is best suited for use as a border, property marker, and in other long run applications. The mortised rail fence offers little protection against the wind and other elements, but it is a good choice for purely decorative and/or boundary needs.

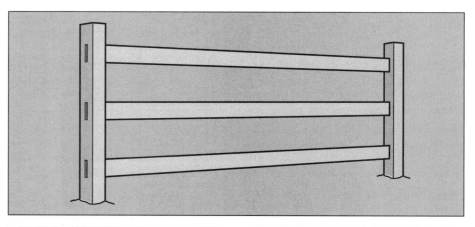

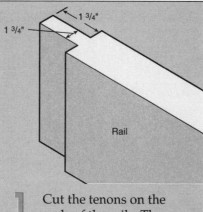

1 Cut the tenons on the ends of the rails. The tenons should be 1 3/4 inches long and 3/4 inch thick. Measure carefully before cutting; the strength of the joint depends on a perfect fit.

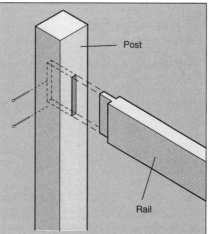

3 When all the mortises have been cut, erect the posts (at 8 feet on center) as described in steps 1 to 9 of "Build the Basic Fence." Place the rail tenons in the mortises, and secure them by driving nails through the side of the posts.

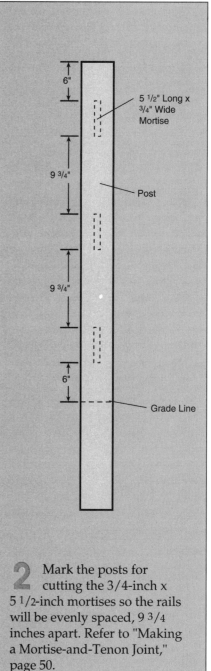

2 Mark the posts for cutting the 3/4-inch x 5 1/2-inch mortises so the rails will be evenly spaced, 9 3/4 inches apart. Refer to "Making a Mortise-and-Tenon Joint," page 50.

Palisade Fence

This fence is unique in that it is made entirely of boards. Each board is self-supporting, so there is no need for posts or rails. Each section of the palisade fence is built as a separate unit, then placed in-ground and anchored with concrete. Adjacent sections are simply nailed together. The major advantage of the palisade is that it can follow the contours of a curving, uneven boundary better than any other type of wood fence.

Materials List

6x4 FENCE SECTION

11	2x6x9 boards
1/4 lb.	10d galvanized nails
5	80-pound bags of premixed concrete

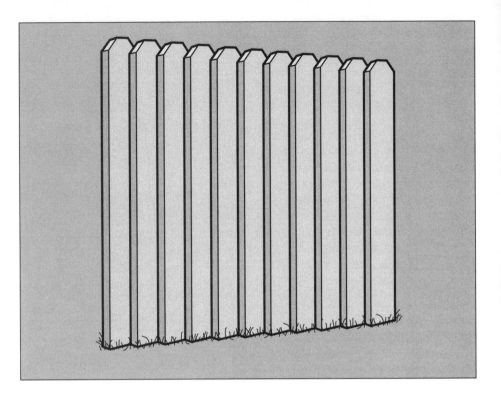

The palisade fence relies on concrete to give it solidity; it is surprisingly strong, despite the absence of any posts or rails.

This fence is equally suited to both long and short runs. Use it to line a curvy driveway or create your own outdoor room or living space. A tall

palisade fence has the advantage of providing three important functions: privacy, security, and a boundary.

1 You may wish to cut dog-ears on the tops of the posts. To do this quickly and easily, clamp several posts together, with the top one marked for cutting. Save the top post to use as a master for cutting additional tops.

Clamp · *Saber Saw* · *45°* · *Boards*

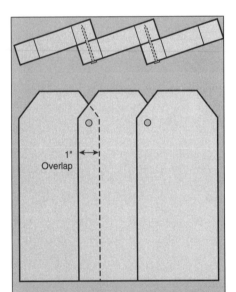

2 When assembling this fence, you must overlap the previous board by 1 inch. Drive a nail in the top and bottom of each board. This 1-inch overlap can be increased or decreased slightly to curve the fence as desired.

1" Overlap

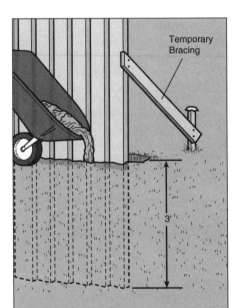

3 Dig a 3-foot deep by 4-foot long trough; it need be only about 6 inches wide. Place the fence section in it, and install temporary bracing as needed. Fill the trough with concrete. Once the concrete has set, the bracing can be removed.

Temporary Bracing · *3'*

DECORATIVE TOUCHES

From an eye-catching post top to imaginative plantings, there are many ways to enhance a fence. This chapter presents various ideas designed to help you get thinking. With a little imagination and some ingenuity, you can add a touch of class to your fence.

Decorative Post Tops

Since the posts are the starting point for building a fence, they are the natural place to begin when attempting to give your fence a custom look. There are two choices here: add prefabricated finials or make your own decorative tops.

Prefabricated finials are ornaments specially designed for topping posts. They are usually made of Douglas fir or hemlock, and are available in many classic styles.

Finials are easy to install. First, drill the appropriate sized screw hole in the post top. Then screw the finial into place.

If you have a measure of expertise working a lathe and other wood turning tools, you can save money and turn your own decorative tops.

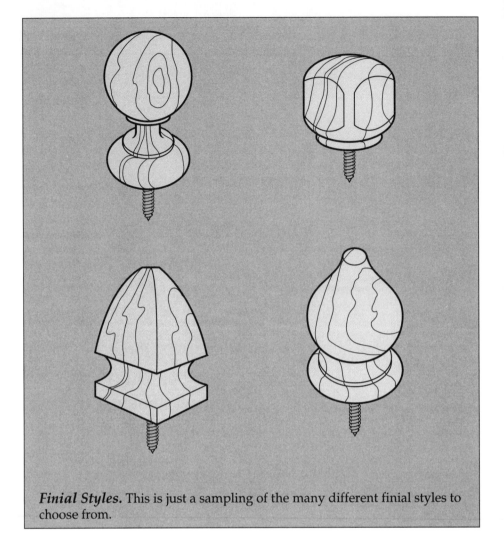

Finial Styles. This is just a sampling of the many different finial styles to choose from.

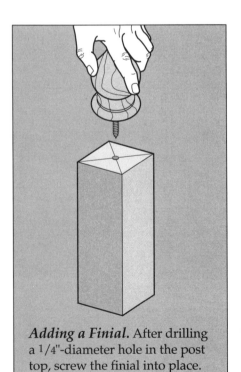

Adding a Finial. After drilling a 1/4"-diameter hole in the post top, screw the finial into place.

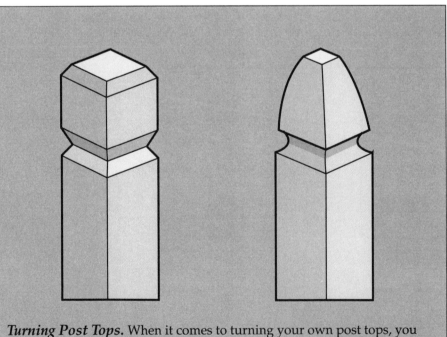

Turning Post Tops. When it comes to turning your own post tops, you are limited only by your imagination.

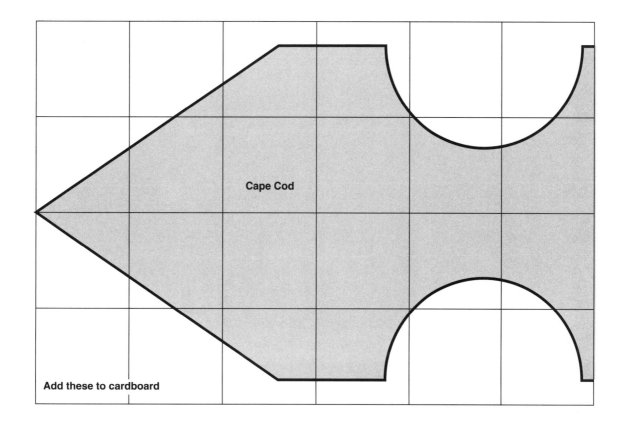

Cape Cod

Add these to cardboard

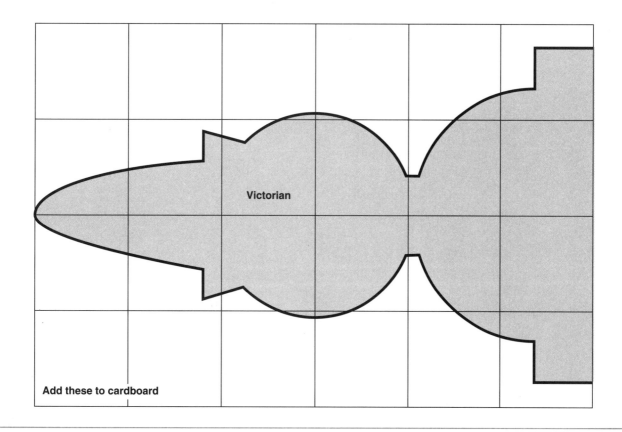

Victorian

Add these to cardboard

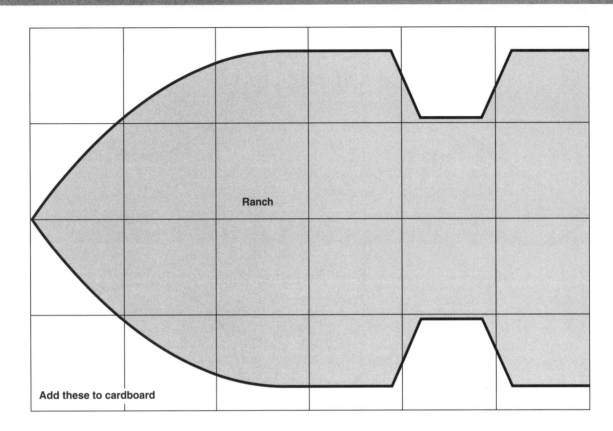

Ranch

Add these to cardboard

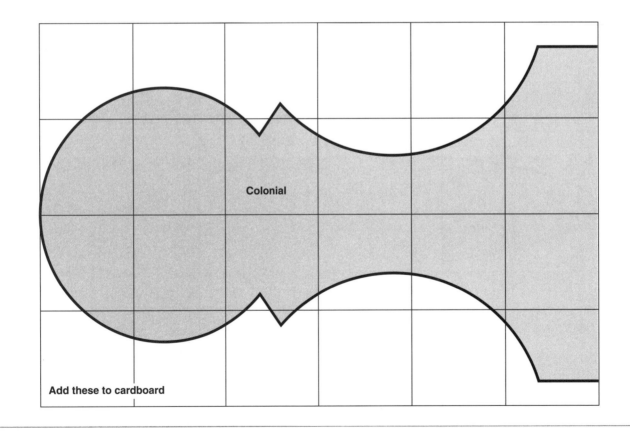

Colonial

Add these to cardboard

Creative Picket Designs

In many areas of the country, it is becoming more difficult to find precut pickets available in different styles. Those lumber dealers that do carry ready-made pickets usually offer a limited choice. You may be able to find a dealer that will cut the pickets from standard stock upon request. In the end, if you plan to build the picket fence on page 29, and you want fancy tops, you may very well end up having to do the job yourself.

1 Transferring Design. Make a template of the design on heavy cardboard, then trace it on the picket.

2 Cutting Pickets. Most portable saber saw blades are narrow enough to cut tight curves. Follow the cutting line by eye, being careful not to force the blade.

3 Cutting Curves. If the curve is straining the blade, either switch to a narrower blade or remove some of the waste with straight cuts.

4 Alternating Styles. An interesting look can be achieved by cutting two different picket top styles, then alternating them on the fence.

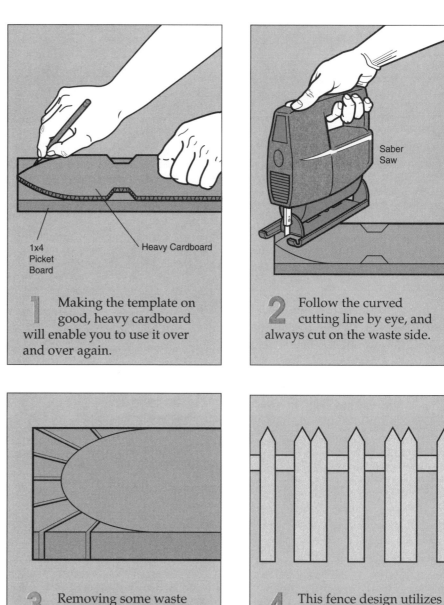

1x4 Picket Board

Heavy Cardboard

Saber Saw

1 Making the template on good, heavy cardboard will enable you to use it over and over again.

2 Follow the curved cutting line by eye, and always cut on the waste side.

3 Removing some waste with straight cuts provides greater clearance for the blade.

4 This fence design utilizes two different picket application styles.

Fence Plantings

An imaginative use of plantings can add a whole new dimension to a fence. Plants soften the harsh lines of a fence, making it a more integrated part of the landscape. Plants can also add new textures, patterns, and color accents to a fence. In turn, the fence acts as a windbreak and provides valuable shelter for shade-loving plants. When choosing plants, you should always consider three factors: the fence itself, the surrounding landscape, and the climate in which you live.

Delicately Constructed Fences. For fences made of bamboo or very thin basketweave, use vines and shrubs to create a light graceful look. Choose thin vines with small leaves—climbing ivy, clematis, and morning glories are good choices. Avoid heavy masses of foliage that can easily blot out a more delicate fence.

Heavily Constructed Fences. For fences made of solid panels or boards, use vines and shrubs with large, thick leaves and bushy masses of foliage. Climbing roses and wisteria work well with heavy fences. A dark, solid fence can be made less imposing by placing low to medium plants in the foreground.

Espaliering. This is a process by which plants, vines, or shrubs are trained to grow flat against a surface, in this case the fence. Informal espaliers are not difficult to achieve. They can provide either a thick, bushy surface covering or a scant tracery—the choice is yours. Good candidates for espaliering include apple and fig trees, English yews, and evergreens. In some cases, plants too tender for exposed areas will thrive when espaliered against a sheltered fence.

Delicate Fences. Climbing ivy makes an interesting contrast to this bamboo fence.

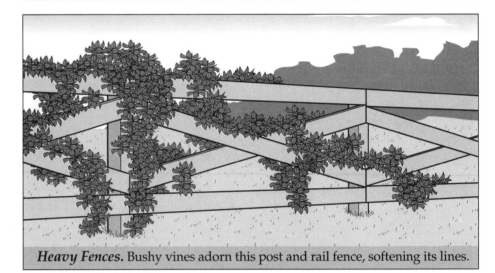

Heavy Fences. Bushy vines adorn this post and rail fence, softening its lines.

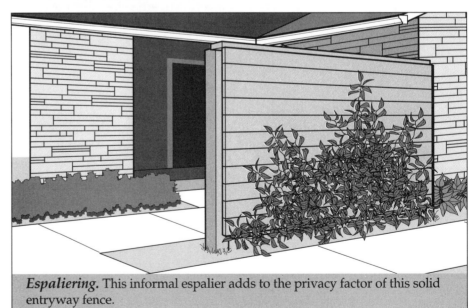

Espaliering. This informal espalier adds to the privacy factor of this solid entryway fence.

BUILD: GATES

A gate is a door, and as such must be planned and designed with the utmost care. The type of fence you are building will influence your choice of gates. But whether simple or ornate, high or low, all gates must meet two requirements: they must look good, and they must work.

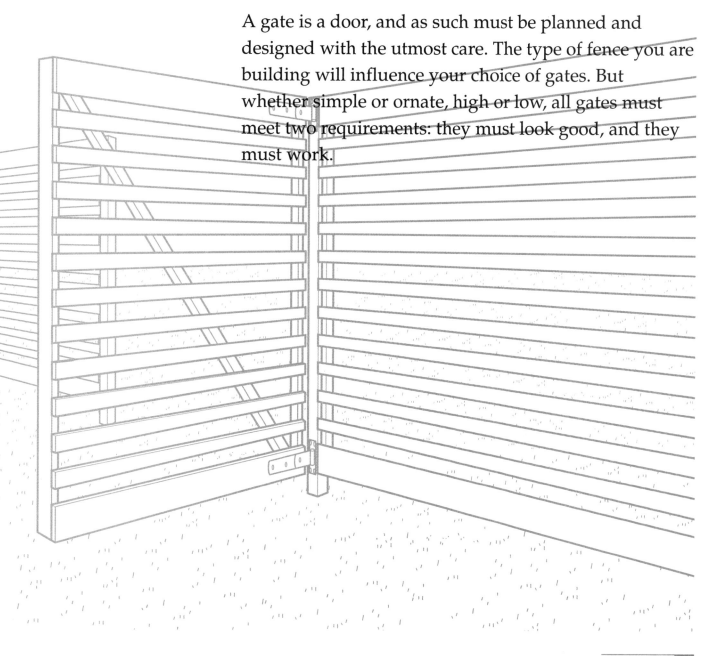

Gate Design and Function

While it is inevitable that the fence will influence the gate, this doesn't mean that the two must be the exact same material and design. Depending on your preference, the gate you build can either complement the fence or stand out from it without creating too sharp a contrast. For example, if the fence is made of vertical pickets, the gate can be made with horizontal or criss-crossed pickets, or the spacing between the pickets can be different, thus creating a pleasing contrast. On the other hand, you may wish to play down the gate so that its position in the fence is hardly noticeable. In this case, incorporate the same materials and design of the fence into the gate. Keep in mind that for some fence styles (the basketweave and both louvered fences, to name a few examples) it will be impossible to match your gate style to the fence. You have to choose either a horizontal or vertical board pattern for those fences.

The height of a gate is determined by the height of the fence. A freestanding gate should be constructed as high as it needs to be for security purposes.

Gates should be made wide enough for two people to pass through, or one person with a wheelbarrow or lawnmower. The distance between gate posts determines actual gate dimensions. A single-section gate should be at least 3 feet wide, but no wider than 4 feet. Gates wider than 4 feet are difficult to support and tend to sag. If the opening you are working with is wider than 4 feet, use a gate with two sections.

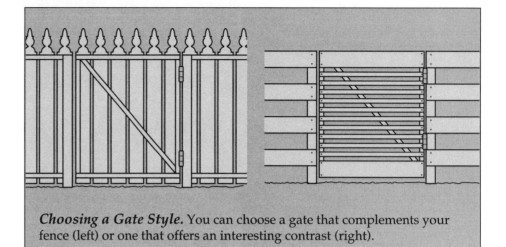

Choosing a Gate Style. You can choose a gate that complements your fence (left) or one that offers an interesting contrast (right).

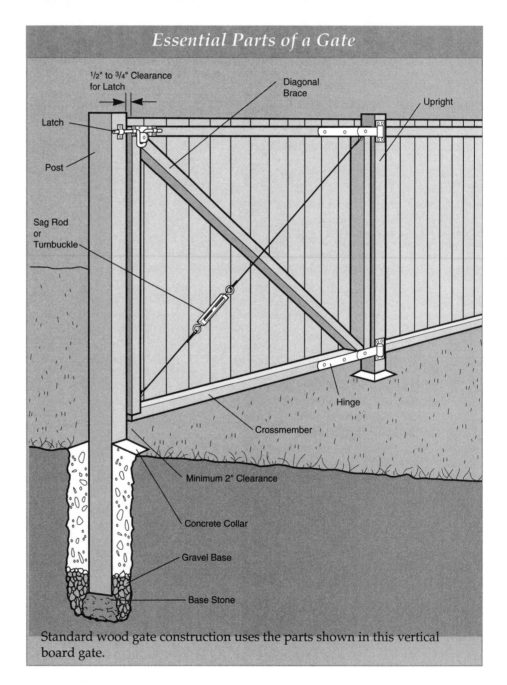

Essential Parts of a Gate

1/2" to 3/4" Clearance for Latch

Diagonal Brace

Upright

Latch

Post

Sag Rod or Turnbuckle

Hinge

Crossmember

Minimum 2" Clearance

Concrete Collar

Gravel Base

Base Stone

Standard wood gate construction uses the parts shown in this vertical board gate.

Gate Hardware

It is best to design and build your gate to accommodate the latches, hinges, and the space between the gate posts.

Latches. The latch keeps the gate closed. Remember that it must withstand its share of rough treatment. A flimsy latch secured with small nails or screws will not last long. Plan adequate side clearance of at least 1/2 inch to accommodate the latch.

Hinges. The main cause of gate failure is inadequate hinges. The hinges must be heavy enough to support the gate's weight and withstand frequent use. Attaching screws or bolts must be large and strong enough to hold the hinges to the gate. Particularly heavy gates may require as many as four hinges. Use hinges with weather-resistant zinc or galvanized coatings. Aluminum or brass hinges are especially durable and rust-resistant. Hinges can also be painted to reduce rust. Plan adequate side clearance of 1/4 inch to accommodate the hinges.

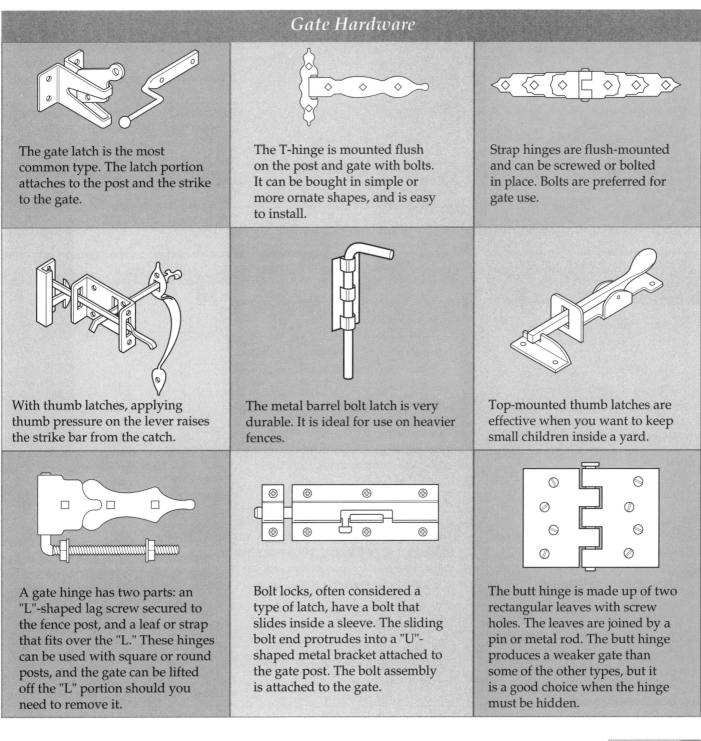

Gate Hardware

The gate latch is the most common type. The latch portion attaches to the post and the strike to the gate.

The T-hinge is mounted flush on the post and gate with bolts. It can be bought in simple or more ornate shapes, and is easy to install.

Strap hinges are flush-mounted and can be screwed or bolted in place. Bolts are preferred for gate use.

With thumb latches, applying thumb pressure on the lever raises the strike bar from the catch.

The metal barrel bolt latch is very durable. It is ideal for use on heavier fences.

Top-mounted thumb latches are effective when you want to keep small children inside a yard.

A gate hinge has two parts: an "L"-shaped lag screw secured to the fence post, and a leaf or strap that fits over the "L." These hinges can be used with square or round posts, and the gate can be lifted off the "L" portion should you need to remove it.

Bolt locks, often considered a type of latch, have a bolt that slides inside a sleeve. The sliding bolt end protrudes into a "U"-shaped metal bracket attached to the gate post. The bolt assembly is attached to the gate.

The butt hinge is made up of two rectangular leaves with screw holes. The leaves are joined by a pin or metal rod. The butt hinge produces a weaker gate than some of the other types, but it is a good choice when the hinge must be hidden.

Sag Rods. Sag rods (also known as turnbuckles or tension supports) are optional for the wood gates described here, but they are easy to install and prevent the gate from sagging. These devices are rods or cables with bolts to secure them to the gate. They should be installed near the top of the upright on the hinge side and run down to the crossmember on the latch side. Use the turnbuckles in the rod to adjust the tension to keep the gate steady. Install the sag rod on the hinge side of the gate. Sag rods can be bought in most hardware stores and building supply stores.

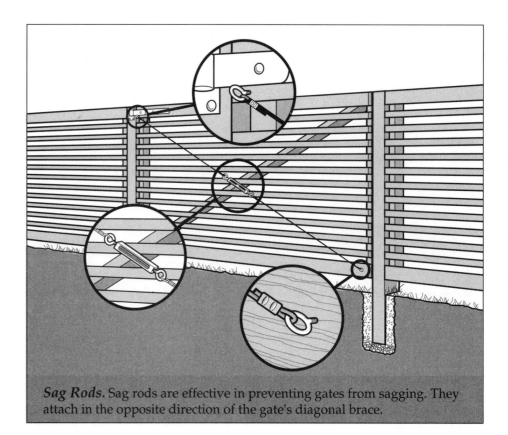

Sag Rods. Sag rods are effective in preventing gates from sagging. They attach in the opposite direction of the gate's diagonal brace.

Securing Gates with Masonry Hinges

Another option is to use special masonry hinges to hang a gate from a brick or block wall. The strap end of the hinge is placed in the mortar joints as the wall is being built. Masonry hinges are ideal for installing heavy wrought-iron gates.

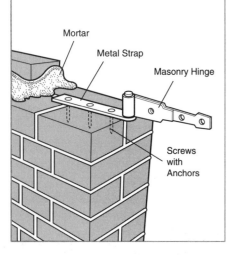

Mortar
Metal Strap
Masonry Hinge
Screws with Anchors

Securing Gates to Masonry with Lead Screw Anchors

Unless your gate is especially large or subject to violent wind gusts, lead screw anchors are sufficient for securing it to a masonry wall. These fasteners are soft lead sleeves that house the wood screw; annular rings add additional resistance to withdrawal from the hole. Because lead screw anchors are threaded internally to match the screw being used, they are quick and easy to install.

1. Select a screw length that equals the thickness of the gate hinge, plus the length of the anchor, plus an additional 1/4 inch. Drill the hole in the masonry the same diameter as the anchor, and 1/4 inch deeper.

2. Insert the anchor into the hole so it is flush with the masonry surface.

3. Mount the hinge, drive the screw into the anchor, and tighten it securely.

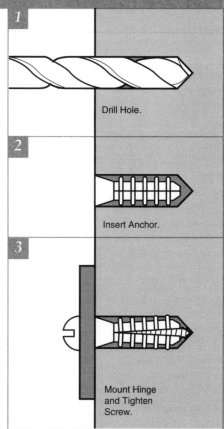

1 Drill Hole.

2 Insert Anchor.

3 Mount Hinge and Tighten Screw.

Making a Lever-Type Latch

A lever-type latch is best suited to more rustic fences, such as the ranch rail or post and rail varieties.

Materials List

1	1x2x12 latch bar	2	3d galvanized barbed nails
1	1x2x6 retainer	7	6d galvanized barbed nails
1	1x2x10 guide	2	10d galvanized barbed nails
3	1x2x2 nailing blocks		
2	1x2x4 nailing blocks		

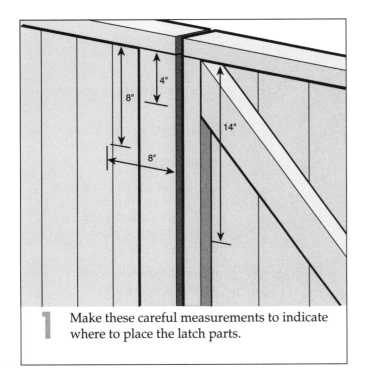

1 Make these careful measurements to indicate where to place the latch parts.

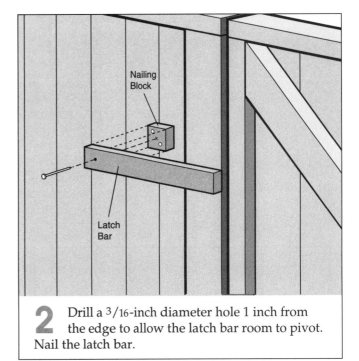

2 Drill a 3/16-inch diameter hole 1 inch from the edge to allow the latch bar room to pivot. Nail the latch bar.

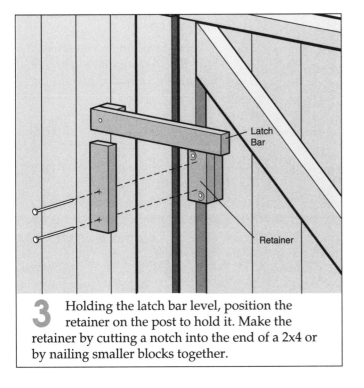

3 Holding the latch bar level, position the retainer on the post to hold it. Make the retainer by cutting a notch into the end of a 2x4 or by nailing smaller blocks together.

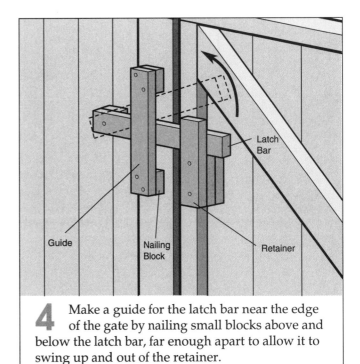

4 Make a guide for the latch bar near the edge of the gate by nailing small blocks above and below the latch bar, far enough apart to allow it to swing up and out of the retainer.

Making a Sliding Latch

A sliding latch is ideal for the more contemporary fence styles, such as the lattice top and alternate panel.

Materials List

3	1x2x6 guides
1	1x2x16 latch bar
1	1/2"-diameter dowel x 4 handle
8	1x2x2 nailing blocks
12	6d galvanized barbed nails

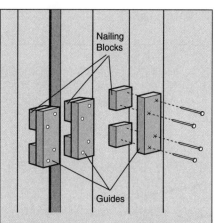

1 Start with 2-inch boards to make the guides, two on the gate and one on the fence post. The guides should be fairly close fitting on the nailing blocks, since all the latch bar has to do is slide in them.

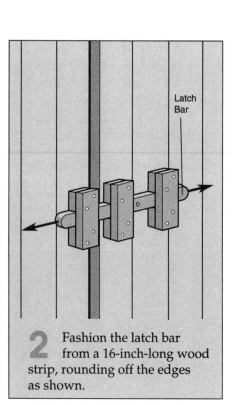

2 Fashion the latch bar from a 16-inch-long wood strip, rounding off the edges as shown.

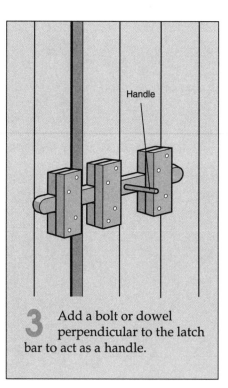

3 Add a bolt or dowel perpendicular to the latch bar to act as a handle.

Open Gate Latch

A latch that keeps the gate propped open can be very useful, especially when your arms are full. To make this type of latch, proceed as follows:

Materials List

1	1x2x12 stake
1	1x2x8 catch
1	1-3/4 carriage bolt and nut
1	2-1/2 carriage bolt and nut

1. Swing the gate wide open, and mark the spot on the ground where you want it to be held.

2. Drive a 1-inch stake, about 12 inches long, into the ground. Leave enough of the stake protruding that it almost touches the bottom of the gate.

3. Use an 8-inch-long piece of wood to make the catch. Approximately 2 inches from one end of the catch, cut a notch wide enough to hold the bottom of the gate. Square the notch off, but taper that end of the wood to a blunt point. Add a heavy nut or other counterweight to the opposite end.

4. Mount the catch to the stake with the 2 1/2-inch bolt, with the tapered end pointing toward the gate.

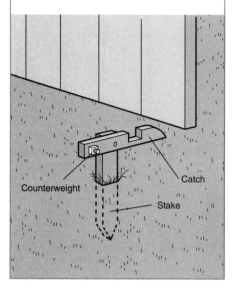

Building a Simple Horizontal Gate

Following is one of the most common gate construction techniques. The gate is built as a separate unit and then hung between the presunk 4x4 fence posts. The frame of the gate is constructed of 2x4 lumber to provide the required strength and durability.

If you build the picket fence, or any other design in which the fence boards are flush with the front of the posts, you will want the front of the gate to be flush with the front of the posts as well. To do so, rip the thickness of the fence boards (usually 3/4 inch) from all the 2x4 gate framework pieces (crossmembers, uprights, and diagonal).

If you build the closed post and board fence, or any other design in which the fence boards protrude out past the front of the posts, you will want the finished board facing of the gate to protrude out past the front of the posts as well. Therefore, you do not have to rip the 2x4 gate framework.

Materials List

2	2x4x42 crossmembers
2	2x4x46 uprights
1	2x4x62 diagonal brace
4	1x6x42 face boards
16	6d galvanized nails
10	#10x3 flat head wood screws

1 Cutting the Crossmembers.
Begin by cutting the two crossmembers. To calculate the length of these pieces, measure the total distance between the posts and subtract 1 1/2-inches. This will allow 3/4-inch clearance on either side of the gate when it is installed. Then cut the two uprights. To calculate the length of these pieces, measure the length of the gate posts (the distance from the ground to the top of the posts). Subtract 14 inches from that length; this allows sufficient clearance at the bottom of the gate and ensures that the board facing will cover the visible framework.

2 Assembling the Framework.
Assemble the gate framework using #10x3-inch flat head wood screws. In this case, the screws provide superior holding power compared to nails. Cut the diagonal brace by measuring the distance from corner to corner on the assembled gate frame. Place this diagonal on the frame where it will be fitted and mark the angles that are required to provide a tight fit. Always attach the lower end of the diagonal on the hinge side of the gate; reversing it will cause the gate to sag. Cut the angles on the diagonal, check the fit, and use #10x3-inch screws to secure the diagonal. Then check the framework to make sure it is square.

3 Attaching the Face Boards.
Use 6d galvanized nails to attach the matching face boards. You will need some temporary bracing or a helper to hold the gate in place as you hang it.

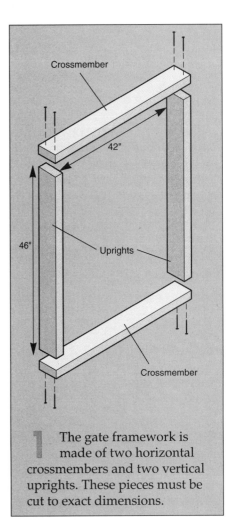

1 The gate framework is made of two horizontal crossmembers and two vertical uprights. These pieces must be cut to exact dimensions.

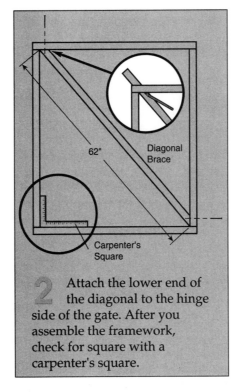

2 Attach the lower end of the diagonal to the hinge side of the gate. After you assemble the framework, check for square with a carpenter's square.

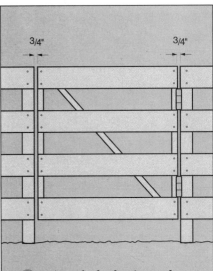

3 Attach the horizontal face boards as shown to match the fence style. Select your hardware and refer to "Hanging a Gate," page 67.

Build a Simple Vertical Gate

The gate is built as a separate unit and then hung on the fence. You can use 1x3 boards for the four uprights and the two crossmembers, and a 2x3 board for the diagonal brace. Wood screws are the first choice for nailing the framework together, but galvanized nails with the ends clinched over will also do the job.

Materials List

2	1x3x42 crossmembers
1	2x3x62 diagonal brace
4	1x3x46 uprights
6	1x4x52 boards
4	#10x2 1/2" wood screws
16	#10x1 1/2" wood screws
24	6d galvanized finishing nails

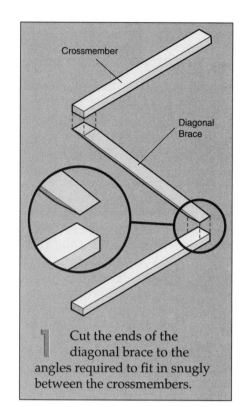

1 Cut the ends of the diagonal brace to the angles required to fit in snugly between the crossmembers.

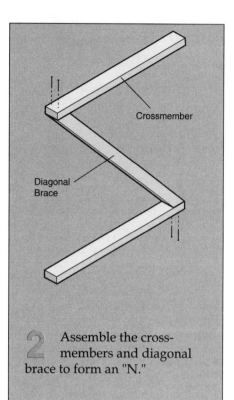

2 Assemble the crossmembers and diagonal brace to form an "N."

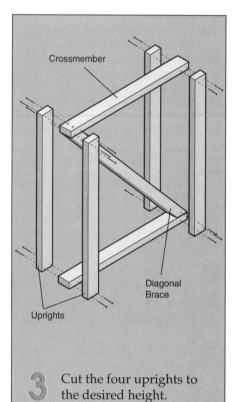

3 Cut the four uprights to the desired height.

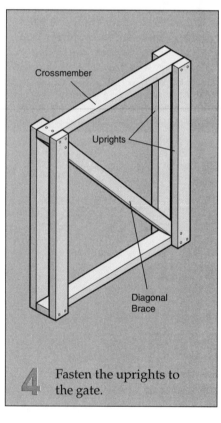

4 Fasten the uprights to the gate.

5 Hang the completed gate as described on page 67.

1 Hanging a Gate

Mounting the Gate. Place the gate on wooden support blocks in the gate opening. Adjust the gate to the desired height as necessary. The gate bottom should be aligned with the fence bottom.

If you want the gate to open inward or the latch is on the inside, align the back of the gate frame with the back of the post. If the gate opens outward or the latch is on the

2 outside, align the front of the gate frame with the front of the post.

Attaching the Hinges. Position the hinges on the gate post and gate and mark the location of the bottom of the top hinge. Mark the screw or bolt holes. Drill pilot guide holes in the post and gate for the screws or bolts. Attach the top hinge. Make sure you leave the support blocks in

3 place under the gate. Repeat the procedure for the bottom hinge.

Attaching the Latch. Position and attach the latch on the gate. Then

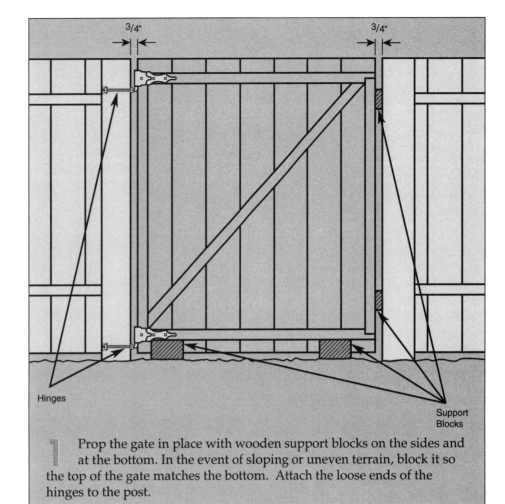

Hinges

Support Blocks

1 Prop the gate in place with wooden support blocks on the sides and at the bottom. In the event of sloping or uneven terrain, block it so the top of the gate matches the bottom. Attach the loose ends of the hinges to the post.

2 Use minimum 2-inch long wood screws or bolts to attach the hinges. Predrill the holes with a drill bit sized for the screws or bolts being used.

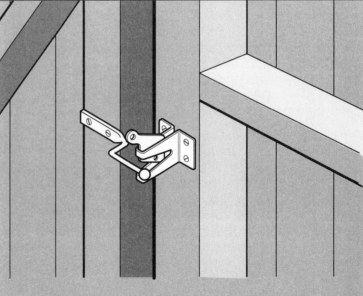

3 Use minimum 2-inch long wood screws or bolts to attach the hinges. Predrill the holes with a drill bit sized for the screws or bolts being used.

Gate Reinforcement

Garden and entry gates normally take alot of abuse. They are exposed to wind and weather, young gate swingers, and hurried pedestrians who push against the framework before the latch is free. Gates hung from fences that are not very rigid, and which tend to expand and contract with the weather, often bind, refuse to latch, and drag on the ground.

In addition to sagging gates (which can be prevented through the use of sag rods, as explained on page 62), loose hinges are another common problem. Loose hinges should be replaced with larger hinges, or the screws replaced with longer ones or even bolts. Be sure to plug the old screw holes before inserting new screws or bolts. Many of the screws that come with hinge sets are too short. Two weak hinges can be strengthened with a third one of similar size, placed a bit above the midpoint between the two.

A leaning post is another major cause of trouble. You may be able to remedy this problem simply by straightening the post and tamping the soil around it. You can also wedge the post tightly into the earth using pressure-treated wedges on each post face. But these may turn out to be only temporary solutions. If the post is not already set in concrete, it probably should be.

A pair of gate posts that won't remain stable can cause the gate to be out of line. They should be supported by tying them together below ground level with concrete.

1 Digging the Trench. Remove the turf or paving between the posts and dig a trench as wide and deep as the post holes. Brush the soil from the footings and tamp the soil in the floor of the trench.

2 Adding a Base of Stone and Gravel. Line the bottom third of the trench with loose stone and with gravel.

3 Filling with Concrete. Shovel concrete into the trench with concrete up to the level required to support the turf or paving.

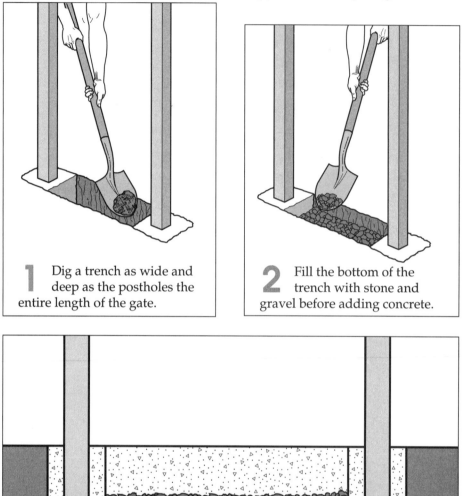

1 Dig a trench as wide and deep as the postholes the entire length of the gate.

2 Fill the bottom of the trench with stone and gravel before adding concrete.

3 The concrete trench will anchor even the most wobbly gate.

PRESERVING & REPAIRING WOOD

Even the best-built wood fences can have problems. The accumulation of dirt, leaves, and moisture leads to wood decay. Shade and damp conditions encourage mildew and fungus. Wood color and texture deteriorate through the weathering effects of sun, rain, freezing, and thawing. These problems can be greatly lessened through sound wood-preserving techniques, as well as regular maintenance and sound repair work.

Wood-Preserving Treatments

Decay is the result of wood-rotting fungi attacking and destroying wood. Outdoor wood near or in contact with the ground is especially susceptible to decay. Decay can be delayed by the application of chemical preservatives. These chemicals can be brushed or rolled on. Wood pieces can also be soaked in them, or the chemicals can be forced into the wood under pressure. This last process is called pressure treating.

Preservatives make possible the use of woods that normally would not be suitable for direct contact with the ground. This means that the less expensive woods such as pine and spruce can be made to last as long as the three naturally rot-resistant woods: heart cedar, cypress, and heart redwood. In these species, the wood from the middle of the ree, called the heartwood, is what is decay resistant. The wood from the outer layer, called the sapwood, is not.

Three of the most popular and effective wood preservatives are creosote, pentachlorophenol, and copper naphthenate. Creosote is known for its sharp, asphalt-like smell. In residential applications, this smell usually prevents its use. But in the country, creosote-treated fences will last many years.

Pentachlorophenol, or penta, has long been used as a preservative. It is available as a ready-mix and is more popular than creosote for residential use because it is cleaner and has only a slight petroleum smell, most of which disappears when the oils evaporate. Unlike creosote, penta can be painted.

Copper naphthenate is usually more expensive than the other two, but it has the advantage of being safe around plants. It is noncorrosive and odorless, and stains the wood a light green. Copper naphthenate does not leach out into the soil, another reason it is good for residential use.

Application Techniques

Soaking wood directly in cold solution is the simplest and least expensive method of applying preservative. Use an oil drum or similar container to treat posts, since only the below-ground portion plus a few inches need to be soaked. To treat whole posts, make a homemade trough using several layers of heavy polyethylene sheets draped between cross ties. The trough should be deep enough to immerse the longest members without having to rotate them. Pine takes about a day to absorb 4 to 5 pounds of pentachlorophenol per cubic foot of wood. Other woods might require as long as a week to soak up the same amount.

The most common method of applying preservative is to brush, roll, or spray it on the wood. If you are patient and willing to apply several coats over consecutive years, the wood will eventually absorb enough preservative to protect it substantially.

When wood not thoroughly saturated with preservative is cut, the ends should be soaked again or brushed thoroughly with preservatives. The

Applying Preservatives. An oil drum is ideal for soaking fence posts in preservatives, since only the below-ground portion plus a few extra inches need to be soaked.

ends are the likeliest places of moisture entry. Any cut or bored hole in the lumber should receive an extra treatment of preservative.

Never forget that the solvents in some preservatives are flammable. If possible, it is best to avoid these types. If you must use them, take precautions against fire. Some preservatives are highly irritating, and direct contact with the skin can be harmful. Wear goggles, rubber gloves, and a respirator when working with preservatives.

Caution: *Take extra care to prevent that children and animals cannot get near the wood while the preservative is curing.*

Brush Application. Applying preservative to a wood fence by brushing normally requires several coats, but the protection it provides is well worth the effort.

Roller Application. A roller will paint a solid board or panel fence in no time at all. Use a clip-on screen to roll out the excess paint.

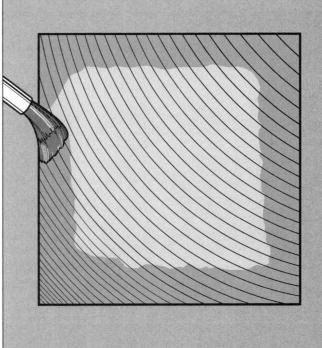

Working with Treated Wood. Brush additional preservative on the cut ends of treated lumber for added protection.

Pressure-Treated Wood

Preservatives are most effective when applied under pressure. These so-called water-borne preservatives are odorless and clean, and wood treated with them can be painted. Pressure-treated lumber is safe for human and animal contact. It can last 50 years or more, even when exposed to moisture and insects.

Another advantage of pressure-treated wood is that it is less costly than any of the naturally rot-resistant woods. Pressure-treated lumber comes in two classifications: LP-2 for aboveground use and LP-22 for below-ground use. A dust mask is an absolute necessity when sanding pressure-treated lumber.

Fence Maintenance

There are many finishes that will protect a fence; these include paint, stain, and clear waterproofing sealers. The only woods that cannot be painted or stained after being treated with preservatives are those with naturally oily surfaces.

Paint

Paint is an excellent wood preservative. It seals and protects wood surfaces. The wood should be clean, dry, and primed before being painted. An oil-based primer is best for raw wood. For a top coat, use a durable, exterior wood-finish paint, either oil-based or latex. Latexes are easier to work with and are longer lasting.

Stain

Film-forming finishes are not recommended for fences. Shellac discolors and cracks when exposed to weather. Varnish can yellow and needs constant renewing. Urethane plastic varnishes are more durable, but they must be completely removed before applying a new coat of varnish.

Stain allows some qualities of the wood to show through, while still protecting the fence against weathering. Semitransparent penetrating stains present a uniform appearance, allowing the wood grain to show through nicely while adding some color. They should be used on new wood and for recoating clean, previously stained wood in good condition. Heavy-bodied or solid-color stains cover the wood grain but not the texture. They are best for older wood that needs to be rejuvenated.

Clear Waterproofing Sealer

Lumber that is not naturally resistant to decay tends to splinter with exposure to weather. (Commonly used woods in this category include spruce, hemlock, birch, hickory, red oak, and poplar.) Use a clear waterproofing sealer for these wood species. These water-resistant coatings are sometimes called repellents. They help prevent rain and moisture from soaking into woods. Apply repellents annually to preserve natural wood color for many years.

Bleach

It is also possible to freshen wood color that has become gray or faded with age. A home-made remedy for freshening wood color is a mixture of one part household liquid bleach to three parts water. This solution should then be scrubbed into the wood with a brush.

Painting. A portable sprayer makes quick work of finishing your fence.

Restoring. Brush on a solution of liquid bleach and water to restore the natural wood color. Wear gloves for protection.

Fence Repair

The job of checking a fence periodically for evidence of damage or wear is an important one. A wobbly fence post is an example of a serious problem that should be corrected immediately. This can be caused by posts that were not properly set or pressure-treated to or because moisture, freezing, and thawing have loosened their buried ends.

▪ If the post is still in fairly good condition, you often can steady a wobble with 2x4 stakes. Bevel the driving ends, soak the stakes in preservative, then drive them on opposite sides of the post and bolt them together.

▪ Concrete makes a more permanent repair. Enlarge the posthole, pour the concrete, and tamp it. Braces will keep the post plumb while you work.

▪ Mend posts broken below the ground with aluminum or wood splints. Enlarge the hole, bolt the splints, and pour the concrete.

▪ If the post has rotted so badly that it must be replaced, either rent a post puller or inch the post out with a long crowbar. Dig away the earth around the posts with a spade to make the removal easier.

Of all the rails, rot usually attacks the bottom one first. By spotting decay early, you can saturate the damage spot with a preservative, then end it. If the rail has broken away, you will have to replace it with treated wood.

▪ Shore up rotted rails with a short length of 2x4. Butt it under and tightly against the rail, and fasten it to the post with galvanized nails.

▪ Damaged rails can also be repaired with T-plates. Drill pilot holes for the screws, and paint the brace to match the fence.

▪ Apply a liberal amount of butyl caulk at the rail/post joint. It will deter rot for several years.

A simple solution to shore up a rotted rail is to nail a 2x4 support block under it.

T-plates provide a stronger method of repair for damaged rails.

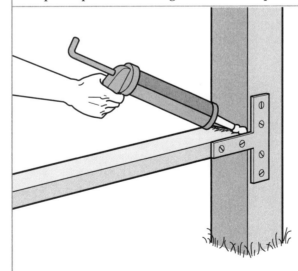

Butyl caulk applied at the rail/post joint is an excellent preventive measure against further rotting.

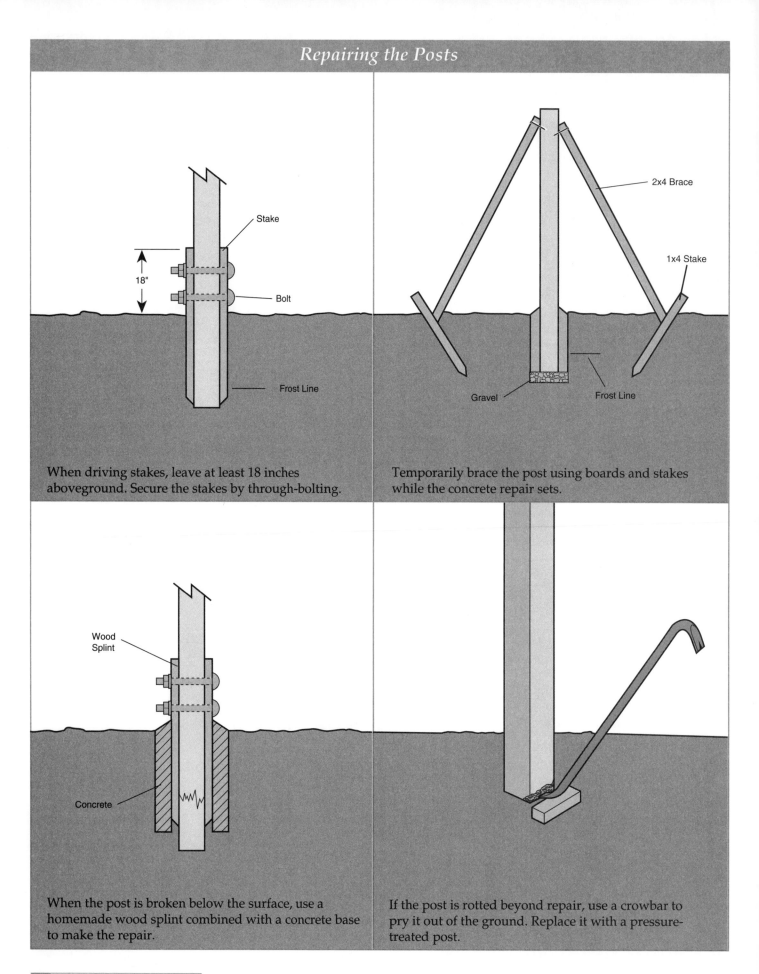

Stake

18"

Bolt

Frost Line

2x4 Brace

1x4 Stake

Gravel

Frost Line

Wood
Splint

Concrete

When driving stakes, leave at least 18 inches aboveground. Secure the stakes by through-bolting.

Temporarily brace the post using boards and stakes while the concrete repair sets.

When the post is broken below the surface, use a homemade wood splint combined with a concrete base to make the repair.

If the post is rotted beyond repair, use a crowbar to pry it out of the ground. Replace it with a pressure-treated post.

OTHER TYPES OF FENCES

When it comes to fencing materials, the choices reach far beyond only wood. Before deciding upon a fencing material, be sure to consider such alternatives as aluminum panel and chain link. Each provides its own unique look and feel.

Aluminum Panel Sectional Fences

Prefabricated aluminum panel sectional fences offer all of the obvious advantages of metal: strength, permanence, and low maintenance.The design choices in aluminum fencing can vary from adaptations of traditional wood fence styles such as picket and louver to the more common ornamental. It is available in a variety of baked enamel colors and finishes. Aluminum fencing comes in the standard 6- and 8-foot widths, and in heights starting at 3 feet and going up to very high sizes.

A prefabricated aluminum fence is designed to be built quickly and easily. It is as suited to long runs as it is to short lengths. An aluminum fence is ideal for surrounding a swimming pool and for bordering home and property. While it offers little shelter from the elements, it is an excellent security fence.

1 There are generally three different types of posts provided with prefab aluminum fencing: line, end, and corner. Line posts have holes punched on opposite sides to accommodate the fence sections, end posts have only one side punched, and corner posts are punched on adjacent sides.

2 If post caps are provided, install them before sinking the posts in the ground. Post holes for aluminum fences should be dug at least 18 inches deep for 36-, 42-, and 48-inch high fences, 24-inches deep for 60-inch high fences, and 30 inches deep for 72-inch high fences. As with other fences, the concrete footing should extend below the frost line. Dig holes for the line posts every 6 or 8 feet along the fence line, depending on the width of the sections.

3 Once the first end post has been sunk and the concrete footing has set, slide the horizontal rails of a fence section into the holes

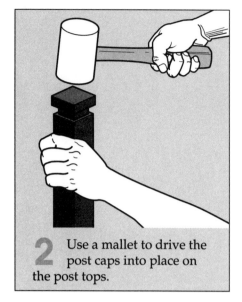

1 Before installing, separate and identify the different posts provided.

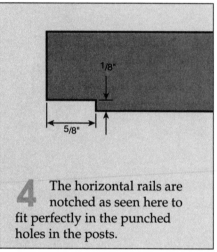

2 Use a mallet to drive the post caps into place on the post tops.

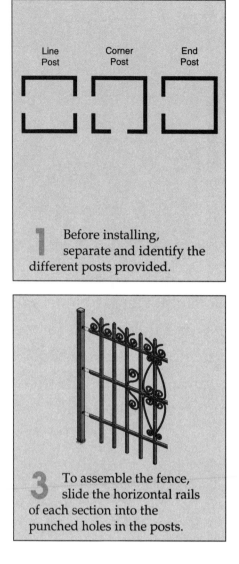

3 To assemble the fence, slide the horizontal rails of each section into the punched holes in the posts.

4 The horizontal rails are notched as seen here to fit perfectly in the punched holes in the posts.

as far as the notched ends of the rails will allow. Place a line post into the next hole, sliding the holes in the line post onto the horizontal rails of the fence section as far as possible. Pour concrete around the line post, tamp it down, and check for plumb and level. Repeat this procedure for all the fence sections.

4 If a section of fence is less than an even 6 or 8 feet, it will be necessary to shorten the section by cutting with a hack saw. Notch the ends of the horizontal rails so they will fit into the post holes.

5 The aluminum fence will follow the contour of the ground because of its ability to rake. On unusually rough ground, you can raise or lower a post so that the line

5 Ornamental aluminum lends an elegant touch to most any residential setting.

of fence is gently rolling rather than a sharp up and down pattern.

Chain Link Fences

If security is your main concern, chain link is one of the most reliable fences you can choose. It has been in widespread use for years as a means of keeping children in yards and pets in runs, and for providing security around swimming pools and other outdoor areas. Most chain link is galvanized, but you can also get it coated with colored vinyl, which has proven effective in preventing rust.

Chain link comes in standard 4-, 6-, and 8-foot heights. When selecting a wire thickness, understand that the smaller the gauge number the thicker (and more expensive) the wire. For most effective security, a mesh opening no larger than 1 3/4 inches is recommended. The typical chain link installation kit comes with everything you need, except the fence puller or come-a-long (which can be rented) and concrete.

Materials List

1	chain link section
2	6-foot posts
1	post cap and line post eye cap or
2	line post eye caps
1	top rail
4	rail tension bands and fittings
10-15	tie wires

1 **Setting the Posts.** Stake out the fence as described in steps 1 to 3 of "Build the Basic Fence." Chain link should be attached to posts no farther apart than 10 feet on center. Whenever possible, make the sections even. For example, a 63-foot long fence should have the posts set 9 feet apart, thus giving you seven equal fence intervals.

2 **Adding the Hardware.** Pour the concrete footings and allow them to cure for several days. Install fittings on the corner and end posts, 1 foot apart with the flat side facing the outside of the fence. Then add the rail tension bands and tap the post caps into place. Place loop caps on the remaining posts, flat side facing out.

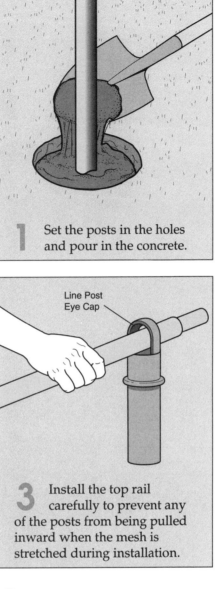

1 Set the posts in the holes and pour in the concrete.

3 Install the top rail carefully to prevent any of the posts from being pulled inward when the mesh is stretched during installation.

3 **Installing the Top Rail.** It is made to fit snugly through the loop caps of the end and corner posts.

4 **Unrolling the Chain Link.** With the framework complete, unroll the chain link along the outside of the frame, provided you have enough room.

5 **Installing the First Tension Bar.** Slip a tension bar through the end of the mesh and onto the rail tension bands. Tighten the bands, then begin moving along the fence and loosely fasten the mesh to the top rail with tie wires.

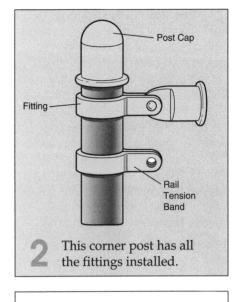

2 This corner post has all the fittings installed.

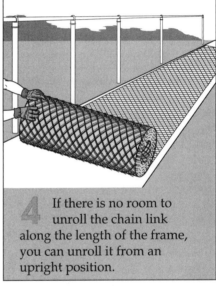

4 If there is no room to unroll the chain link along the length of the frame, you can unroll it from an upright position.

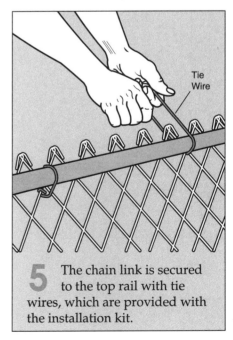

5 The chain link is secured to the top rail with tie wires, which are provided with the installation kit.

6 Positioning the Second Tension Bar. Install a second tension bar approximately 9 or 10 feet from the end post, and attach the fence puller or come-a-long to it. If there is too much slack in the chain of the pulling device, place the operating lever in the neutral position and "freewheel" until the chain becomes taut. Take up the slack in the fence slowly, avoiding any jerky movements. If there is too much slack in the fence, you will have to reposition the tension bar.

7 Tightening the Fence. Shake the chain link and give the puller another couple of turns to get the mesh extra tight. Tighten this set of rail tension bands, as was done on the first post. Continue this procedure for the remainder of the fence installation.

8 Adding the Gate. Attachment of the gate is the final step in the procedure. Bolt the support hardware to the gate post, then hang the gate on the hinge pins. Adjust the gate latch as necessary.

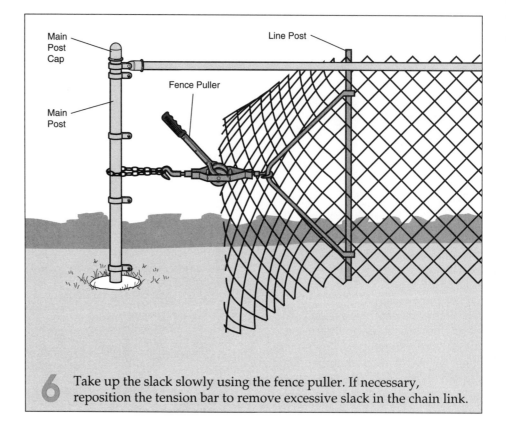

6 Take up the slack slowly using the fence puller. If necessary, reposition the tension bar to remove excessive slack in the chain link.

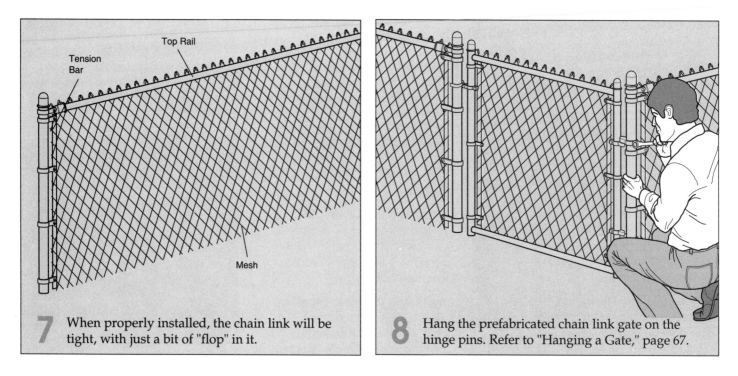

7 When properly installed, the chain link will be tight, with just a bit of "flop" in it.

8 Hang the prefabricated chain link gate on the hinge pins. Refer to "Hanging a Gate," page 67.

Auger A hand- or machine-operated tool with a screwlike shank for boring holes in soil.

Brace Often used temporarily, an inclined piece of framing lumber used to make a post or fence section rigid.

Butt joint Two pieces of wood joined by attaching the square-cut end of one member to the end or face of another.

Butyl caulk Caulk with excellent sealing qualities that can be used at the post/rail joint to help deter rot.

Chain link Very strong metal fence material manufactured with interlocking wire mesh.

Chalk line A string or cord covered with chalk that is snapped against wood members to make a mark for measurement or cutting.

Clear finish Wood finishes that allow the wood grain to be seen.

Cleat A short length of wood fastened to a post or other structural member to provide support.

Clinch To drive nails through wood and bend the points down flat on the other side.

Concrete collar A ring of concrete placed around posts to securely anchor them; also prevents water from collecting around the posts.

Curing The slow chemical action that hardens concrete.

Decking nail Nail with a spiral or screw-type shank that resists loosening over time.

Dado joint A joint in which one board fits into a rectangular groove cut across the grain of a board.

Espaliering The process by which plants, vines, or shrubs are trained to grow flat against a surface, for example a fence.

Fence bracket Connector used in combination with nails to secure rails to fence posts; eliminates the need for special cutting, notching, and fitting.

Fence puller Device used to stretch wire fencing before it is permanently fastened to the posts.

Finial Prefabricated ornament used to top off a fence post.

Floodgates A water gate that spans a stream when a fence line crosses water; they permit water and debris to pass, but animals are stopped.

Footing The concrete base supporting posts and other structural members.

Frost heave Movement of the soil and objects in it due to the freezing and thawing process; common problem in cold climates.

Frost line The level below grade beneath which the ground does not freeze.

Galvanizing Coating a metal with a thin layer of zinc to prevent rust. Fasteners and connectors must be galvanized for outdoor use.

Heartwood Wood that comes from the center of the tree; it is darker in color, denser, and more durable than the surrounding sapwood.

Lap joint A joint in which two pieces are lapped one atop the other.

Lattice Open structure of crossed strips of wood.

Lead screw anchor Internally threaded fastener that houses wood screws; used to secure gates to masonry walls.

Louver Fence slats positioned at an angle to promote air circulation and light penetration; may be either movable or fixed.

Mortise A rectangular cavity cut into a piece of wood into which a tenon is inserted.

Nominal dimensions The identifying dimensions of a piece of lumber (e.g., 2x4) which are larger than the actual dimensions (1 1/2" x 3 1/2").

On center A point of reference for measuring. "Eight feet on center" means 8 feet from the center of one post, for example, to the center of the next post.

Plumb On a straight vertical. A fence that is plumb is straight up and down with no lean. When constructing a fence,

plumb must be checked regularly to ensure vertical stability.

Plumb bob A conical weight suspended on the end of a string to test vertical lines.

Post Vertical framing member (e.g., a 4x4 or 6x6) that acts as the foundation of a fence.

Preservative Any substance that will prevent damage from wood-destroying moisture, fungus, or insects.

Pressure-treated lumber Wood treated with chemical preservatives to make it resistant to the harmful effects of decay, weather, and insects.

Rail Horizontal framing member, usually a 2x4 or 2x6, to which the fence boards are attached.

Sag rod Rod or cable installed diagonally on a gate to prevent it from sagging.

Slat Long, narrow strip of wood (e.g., 1x3) used as a substitute for conventional fence rails for the purpose of letting in sunlight and breeze.

Stain Any of various forms of water-, latex-, or oil-based coloring agents, transparent or opaque, designed to penetrate the surface of wood.

Tamp To compact soil, fill material, or concrete with repeated light blows. Concrete is tamped immediately after pouring to remove air pockets and distribute aggregate throughout the mixture.

Template A pattern or guide used to check dimensions, or contours on work that will be replicated.

Tenon A projection cut on the end of a piece of wood that fits into a mortise.

Toenail To nail two pieces of wood together by driving nails at an angle through the edge of one into another.

Windbreak A fence constructed to lessen the force of the wind.